FIRST LEGO® LEAGUE
the unofficial guide

james floyd **kelly** and jonathan **daudelin**

no starch press

san francisco

12 11 10 09 08   1 2 3 4 5 6 7 8 9

ISBN-10: 1-59327-185-9
ISBN-13: 978-1-59327-185-5

Publisher: William Pollock
Production Editor: Megan Dunchak
Cover Design: Octopod Studios
Developmental Editor: William Pollock
Copyeditor: Laura Quilling
Compositors: Riley Hoffman and Kathleen Mish
Proofreader: Cristina Chan
Indexer: Karin Arrigioni

For information on book distributors or translations, please contact No Starch Press, Inc. directly:

No Starch Press, Inc.
555 De Haro Street, Suite 250, San Francisco, CA 94107
phone: 415.863.9900; fax: 415.863.9950; info@nostarch.com; www.nostarch.com

Library of Congress Cataloging-in-Publication Data

Kelly, James Floyd.
  First lego league : the unofficial guide / James Floyd Kelly and Jonathan Daudelin.
      p. cm.
  Includes index.
  ISBN-13: 978-1-59327-185-5
  ISBN-10: 1-59327-185-9
 1. Robots. 2. LEGO toys. I. Daudelin, Jonathan. II. Title.
  TJ211.2.K45 2008
  629.8'92079--dc22
                                                  2008030628

For Decker—I'm sorry, but you're going to get blamed for
all our mischief for a while. . . . Dad

For my parents, Douglas and Vickey Daudelin—Thank you
for your love, guidance, and support. . . . Your son

BRIEF CONTENTS

CONTENTS IN DETAIL

3
GUIDELINES AND RULES   23

4
FINDING EQUIPMENT, MENTORS, AND FUNDING   31

5
STARTING OR BUILDING A TEAM   39

10
BASIC BUILDING 95

11
BUILDING TECHNIQUES FOR THE ROBOT GAME 121

15
THE PROJECT

16
TOURNAMENTS AND BEYOND

Greetings:

Since 1998, *FIRST* LEGO® League (FLL) has invited students aged 9 through 14 in the United States and Canada and up to age 16 globally to participate in an exciting and challenging global competition that encourages both investigation and invention to solve some of the world's most pressing problems.

FLL participants and graduates discover in themselves not only a growing desire to learn but also a drive to apply that knowledge in a positive manner that will benefit the world. I've said many times before that the students involved in FLL are the ones that will find the cures for our diseases, discover alternative fuel sources, and implement food and water shortage solutions—lofty goals, certainly, but all solvable by the right people with the proper skills and motivation.

FLL has over 100,000 students in more than 30 countries and is continuing to grow. Young people around the world are taking positive action now, via their education, that will impact your future and mine.

Please join me in continuing to support the students, teachers, coaches, mentors, and organizations that make FLL such a success.

Sincerely,

Dean Kamen
FIRST Founder

ACKNOWLEDGMENTS

The book you're holding would never have been possible without the hard work of the team at No Starch Press. Bill Pollock, the founder of No Starch and developmental editor for this book, made sure our chapters were coherent and well organized. Megan Dunchak, our production manager, did a wonderful job keeping everything moving as planned, as well as helping with editing. Riley Hoffman made our pictures, screen captures, and graphics presentable. We'd also like to give a special acknowledgment to Derek Yee for creating the nice cover images for our book.

A big thank you also goes to our families and friends who helped and supported the project. Jonathan's parents helped review chapters and provide information from their experiences (Jonathan's mom is an FLL coach). Jim is grateful for the help provided by the talented Atlanta LEGO® MINDSTORMS® network, including LEGO Education representative Kristie Brown, Mischa Holt (*http://www.yesgeorgia.com/*), Mary Roberts, Rick Folea (*http://www.forsythfll.com/*), Rayshun Dorsey (*http://www.wizkidztech.org/*), and Jeff Rosen.

We'd also like to acknowledge our fellow contributors to The NXT STEP blog (*http://thenxtstep.com/*). They were quick to help whenever we needed opinions on many topics and always provided encouragement to keep moving forward. A special salute goes to Dave Parker, who was especially helpful with providing some of the pictures for the book.

Gratitude is also extended to Dean Kamen and everyone at the FIRST organization. We'd especially like to thank Anna Maenhout, Ernie DiCicco, Kim Martineau, and Noriko Morin for their assistance.

Finally, we'd like to thank all the FLL members and teams out there who sent us comments about their experiences in FLL and have participated in our blog's discussion forums. You'll find their advice and tips scattered throughout the book.

INTRODUCTION

If we were asked to describe FIRST LEGO League (FLL) in 50 words or less, here's what we'd say:

> FIRST LEGO League is a fun, challenging, fast-paced, competitive, enriching, and extremely rewarding international competition. Thousands of teams compete to solve problems, build and program LEGO robots, and conduct research and present their findings. FIRST LEGO League proves that math and science can be a lot of fun.

But, of course, there's a lot more to the FIRST LEGO League experience, as you'll learn in *FIRST LEGO League: The Unofficial Guide*. Whether you're completely unfamiliar with FLL, a member of a rookie team, a coach, a parent, or a mentor, you'll find the information you'll need to make real progress during your first competition season, whether it's building and programming robots or performing the research to present to a panel of judges.

In the course of writing this book, we interviewed dozens of teams around the world, including many award winners, and collected a wealth of information that both rookie and veteran teams will find useful. You'll find real-world tips from coaches, students, and judges scattered throughout the book that your FLL team can immediately put to use.

How This Book Is Organized

Here's a look at what you'll find in *FIRST LEGO League: The Unofficial Guide*.

Chapter 1, "What Is FIRST LEGO League?" provides background on the FIRST organization and the four competitions it supports: FIRST Robotics Competition (FRC), FIRST Tech Challenge (FTC), FIRST LEGO League (FLL), and Junior FIRST LEGO League (JFLL).

Chapter 2, "How FLL Works," discusses the four specific components of an FLL competition—the Robot Game, Robot Design, Project, and Teamwork. Each of the four components is covered in more detail in later chapters.

Chapter 3, "Guidelines and Rules," focuses on the guidelines and rules of FLL and the importance of understanding them thoroughly. Examples from previous FLL missions are covered.

Chapter 4, "Finding Equipment, Mentors, and Funding," covers the three major resources your team will need in order to compete: equipment, personnel, and money. You'll also find sample documents to help your team develop a list of necessary resources and manage its inventory and funding.

Chapter 5, "Starting or Building a Team," covers recruiting and selecting team members and offers suggestions on building a team from scratch, approaching sponsors, and getting support from team members' parents. Veteran teams will also find this chapter helpful for tips on recruiting new members.

Chapter 6, "Managing Your Team," provides suggestions for students participating in a competition, including a discussion of various roles for participants and team goals. We also discuss decision-making and offer suggestions on how to build team cohesion and reduce conflict.

Chapter 7, "The Team Experience," includes summaries of the traits of a good team. Quotes are included from successful winners of past FLL competitions, both students and coaches, as well as ways for the team to share its experience.

Chapter 8, "Coaching a Team," discusses the coach's main objectives, including when and how to provide assistance to the team. Software recommendations are provided to help the coach manage the team's meetings, travel, and competitions.

Chapter 9, "NXT vs. RIS," discusses the two systems used in FLL: NXT and RIS. We discuss both kits, and offer pros and cons for using one system over the other.

Chapter 10, "Basic Building," demonstrates some general techniques for building with the NXT kit. It introduces a building methodology along with suggestions for strengthening a robot and adding flexibility to its shape and design.

Chapter 11, "Building Techniques for the Robot Game," introduces building techniques that are specific to FLL, such as using guide attachments and aiming jigs. It also discusses a general design structure for robots that uses a chassis, a bay, and several attachments.

Chapter 12, "Sensors," includes a thorough discussion of the sensors allowed in competition and how to properly use them for maximum efficiency.

Chapter 13, "Getting Organized for Programming," includes a short discussion on file-naming conventions, saving and backing up data, and program maintenance. Flowcharts are covered as a way to help students determine the proper course of action for a robot before they do any actual programming.

Chapter 14, "NXT-G Programming Concepts," offers a detailed discussion of programming techniques, including looping, case/switch use, and program menu systems. The chapter also covers refining programs to reduce size and complexity as well as more advanced actions using the sensors.

Chapter 15, "The Project," discusses the Project component of FLL. It gives advice on researching information, creating a five-minute presentation, and taking action in the community in the area of your research.

Chapter 16, "Tournaments and Beyond," gives a rundown of typical events that happen at tournaments, plus advice on preparing for them. You'll find discussions of what to expect from the judging sessions, what the judges are looking for, and ideas on how teams can increase their scores. You'll also find suggestions for what to do once the season ends, and how to prepare for next season.

The appendix, "Resources," provides a list of online resources for further exploration and information.

We've also created a special discussion forum for the book where you can submit questions and tips and discuss FLL with the authors and with teams from around the world. Just visit *http://thenxtstep.com/smf/* and click the Book Discussions section to get started.

We hope that you and your team will find some useful information in the pages of *FIRST LEGO League: The Unofficial Guide* that will inspire, help, and enhance your FLL experience.

Now, let's get started and have some fun!

1

WHAT IS FIRST LEGO LEAGUE?

What better way to start a book on FIRST LEGO
League (FLL) than with a chapter that tries to answer
the question, "What is FIRST LEGO League?"

Now, we guess that you didn't really find this book by accident, but instead
are a student, parent, or coach involved in FLL and want more information
on any of the numerous topics related to FLL—robots, programming, judging,
and more. If that's the case, please feel free to jump back to the Contents,
find the chapter that best fits the information you need, and start reading.

Or maybe you really have no idea what FLL is. Maybe the book fell off
the shelf as you walked by it at your local bookstore and the cover completely
grabbed your attention. More likely, though, your son or daughter handed
you a copy and said, "Mom/Dad, I want to start a team at my school next year.
Can you help?" (Another likely scenario is a call from your child's teacher
asking if you would be willing to coach or mentor an FLL team.) If this paragraph describes you, you found the right book.

Whatever your experience with FLL, you'll find tons of useful information that will help increase your chances of winning at competitions and many ways to have more fun getting there. And remember, it's not always about winning when it comes to FLL; it's about the experience of working with friends, making new ones, and learning about yourself and the world around you.

FIRST and the FIRST Robotics Competition

FIRST, a nonprofit organization, was founded in 1989 by Dean Kamen, an acclaimed inventor and entrepreneur best known for inventions such as the HomeChoice portable dialysis machine, the iBOT 4000 Mobility System (a wheelchair that can go up staircases!), and the Segway Personal Transporter (Figure 1-1), one of the coolest transportation devices around.

Figure 1-1: The Segway Personal Transporter was created by Dean Kamen.

The acronym *FIRST* stands for *For Inspiration and Recognition of Science and Technology*, and the title says a lot. In Kamen's own words, his goal in establishing FIRST was "To transform our culture by creating a world where science and technology are celebrated and where young people dream of becoming science and technology heroes."

In a world where school sports and athletes get so much attention (and funding), FIRST is a forum in which teams of academically talented students can display their skills, a competition that pushes students' science, math, and creativity skills to the limits. The FIRST organization was founded in 1989 and began preparations for what became the FIRST Robotics Competition (FRC).

The FIRST Robotics Competition held its inaugural competition in 1992 in a high school gym in New Hampshire, with 28 high school teams and their homemade robot gladiators.

Photo courtesy of Forsyth Alliance/Rick Folea

Figure 1-2: FIRST Robotics Competition brings out some very large robots.

FIRST Robotics Competition

FIRST Robotics Competition is still going strong more than 15 years later and has introduced tens of thousands of high school students to the world of engineering. But FRC isn't the only game in town, so to speak. FIRST has also introduced three other robotic competitions: FIRST Tech Challenge (FTC; formerly FIRST VEX Challenge), FIRST LEGO League (FLL), and Junior FIRST LEGO League (JFLL).

FIRST Tech Challenge (Formerly FIRST VEX Challenge)

Because the FIRST Robotics Competition was such a success, a midlevel event for high school–age students, initially called the *FIRST VEX Challenge* and now the *FIRST Tech Challenge*, was added in 2005. FRC has rules and places some limits on the robot hardware that teams can use, but high school teams still have a lot of freedom. However, many teams lacked the skill to compete, and the cost of producing competitive robots was extremely high. Many schools could not compete either because of the cost or because they didn't have access to the hardware, software, or skills needed to build a competition robot.

FIRST VEX Challenge

The name *VEX* was included to reference the name of the product that the teams used to build their robots; the kit came with wheels, gears, metal beams, and more—all of which can be used to build a competition robot. Students used a version of the C programming language to provide the robots with instructions for completing various tasks in the competition. (While VEX was originally sold as a kit by RadioShack, the company sold the name and rights to the kit to Innovation First, Inc., in 2006.)

A good degree of balance was achieved by requiring teams to purchase the same basic VEX components to build the robots and by limiting the software they could use to program them. The competition was designed in such a way that teams could purchase a single robot kit and have a reasonable chance of winning the competition. As a result, new and less-experienced teams were able to compete at lower costs.

The very first FVC competition took place in Atlanta, Georgia, with 130 teams competing. In 2007, FIRST changed the name of the competition to FIRST Tech Challenge. A new FTC kit uses a LEGO MINDSTORMS NXT Brick as the microcontroller and a mixture of aluminum beams, high-power motors, and standard LEGO pieces to build some rather advanced robots (see Figure 1-3 for an example of an FTC kit robot).

Photo courtesy of Dave Parker

Figure 1-3: FIRST Tech Challenge has its own unique robots too.

FIRST LEGO League

FIRST LEGO League was born in 1999. FLL differs from FRC and FTC in two important ways:

- Only students between the ages of 9 and 14 can compete (in the United States).
- Teams consist of no more than 10 students (it's the same with FTC, but FRC can have up to 20 students per team).

Teams sign up during the year, typically when school starts. Schools frequently have robot clubs that form teams, but homeschool organizations also form many teams.

To join the competition, teams pay a fee to FIRST to purchase the basic FLL competition kit that contains a special mat and LEGO pieces, such as those seen in Figure 1-4, which form the mission models.

Photo courtesy of Maureen Reilly

Figure 1-4: FIRST LEGO League uses LEGO pieces for robots and challenge components.

Themes

Each FLL competition has a theme. For example, the theme for the 2007–2008 season was Power Puzzle. Themes from earlier seasons include Mission Mars, Ocean Odyssey, and Nano Quest. The theme is intended to define the goals for each team. In the Power Puzzle season, for example, one of the robot challenges consisted of the robot harvesting corn from a farm (the farm was drawn on the mat as a cornfield and a barn building, and the corn was a small LEGO assembly made of individual pieces that students put together from the FLL kit). The robot needed to deliver the corn to various locations on the mat for points. Teams also had to develop a Project to present to a team of judges. This presentation must follow the theme, so teams during the 2007–2008 season gave presentations on topics such as solar-powered vehicles, alternative fuel sources, and wind power. Teams anxiously wait to hear next season's theme, and when the new theme is announced, many teams are already hard at work on at least the Project portion of the competition.

The FLL Robot Game

If you ask most students, they'll tell you that the Robot Game round is their favorite part of the FLL competition. In a nutshell, it consists of a playing field on which the robot must attempt to complete a number of missions to score points. Each team competes on the same size field and completes the same missions within a time limit.

Teams are allowed to have two members at the table to interact with the robot (such as switching out components, called *attachments*). They score points based on missions accomplished and penalties assessed (if, for example, they need to rescue the robot if it gets into trouble).

Technical Interview

At the tournament itself, students meet with a group of technical judges who ask them about the design and programming of their robot. Based on this interview, the judges award points for innovation, having a strong grasp of technical concepts, and so on. It is a great opportunity for teams to shine as they explain how they tackled the challenges of building and programming a robot (the judges are there to encourage the students to talk about their work and what they learned in the process, not to critique their work).

Project

Teams are required to put together a presentation (similar to a science fair exhibit) on a topic related to the competition theme. Each team chooses a topic related to the theme, researches that topic, shares their research with the community, and presents its findings to a collection of judges.

Teamwork

Another important way to score points is in the Teamwork-judging portion of the competition. In this often-overlooked opportunity, each team meets with judges who ask how the team members worked together, and the judges give each team a simple assignment that demonstrates how the members work. In addition, they watch the teams throughout the tournament (Robot Game, technical interview, and Project presentation) to see how well the members collaborate and how they interact with other teams. The judges always look for teams that exhibit professional and courteous behavior, and they reward it appropriately.

More on FLL to Come

Since FLL is the topic for this book, we delve much deeper into the four areas of competition in later chapters. For now, though, this should give you a fairly good idea of how FLL works and why it continues to be so successful.

Junior FIRST LEGO League

In 2006, Junior FIRST LEGO League started for students between the ages of six and nine, and it's just as much fun as FLL. JFLL's goals are similar to those of FLL: to encourage young students to learn and experiment. And, of course, to prepare them for FLL.

JFLL uses the same theme as FLL, but it has its own competition rules, and teams create a *Show Me poster* that lets them demonstrate what they learned and the paths they took to complete their goals.

Participation is growing every year, and it's expected that JFLL will continue to grow as more schools, teachers, and parents become aware of the benefits it provides to young students. JFLL is definitely a program to watch as well as encourage.

Where Does LEGO MINDSTORMS Fit In?

Most people are probably familiar with the LEGO Group. Children around the world enjoy snapping LEGO bricks together to build houses, cars, and animals. The company built its reputation on providing products with which children can easily build, disassemble, and, most important, mix together to form new products. Found in numerous colors, LEGO pieces are probably some of the most recognized items in the world.

It should come as no surprise, then, that when LEGO decided to investigate the idea of creating a robotics kit that would allow children (and adults) the ability to mix and match standard LEGO pieces with specialized electronics parts, the final product was a huge success.

LEGO released the LEGO MINDSTORMS RCX kit in 1998, and rumor has it that more than one million of the original kits have been sold. Eight years later, LEGO released the next version, called *NXT*, and the new robotics kit is demonstrating even more success than its predecessor.

Figure 1-5 shows two robots used during the Power Puzzle season. Notice that the robots are not identical; each team can choose the components for its robot from the LEGO MINDSTORMS kits. And the best part? Any LEGO part that's ever been created has the potential to show up in a MINDSTORMS Robot Game because all LEGO components easily connect to one another.

We cover MINDSTORMS in more detail in Chapter 9, but teams can use either the LEGO MINDSTORMS NXT or the LEGO MINDSTORMS RCX systems. As we already mentioned, the RCX system was released first in 1998. The NXT was released in late 2006, and it offers many upgrades including more memory and the ability to use more motors in models.

Photo courtesy of Dave Parker

Figure 1-5: Two LEGO robots, RCX on the left and NXT on the right

MINDSTORMS Components

Teams build the robots using LEGO MINDSTORMS robotics kits that consist of the brains of the robot (called the *Brick*), motors, and several sensors for helping a robot interpret its surroundings. In addition to the electronics, the kit also includes various components such as axles, beams, and connectors that the teams use to build the physical robot and give it strength. We cover building robots in detail in later chapters, where you can find plenty of information related to assembling a robot.

The MINDSTORMS Programming Software

Each type of MINDSTORMS kit comes with software that allows the robot builders to program the robots with autonomous behavior. The programming allows the robot to navigate the mat and complete the missions. Programming robots is covered in Chapters 13 and 14.

What Is the LEGO Group's Involvement?

LEGO makes the MINDSTORMS robotics kit and sells a retail version for anyone to purchase, which is typically referred to as the *Retail Kit*. But there's also a different version of the product called the *Education Kit*. Schools can purchase the Education Kit directly from LEGO Education, a division of the LEGO Group that focuses on using their products in educational markets.

What Are the Differences Between the Two Versions?

The Retail Kit comes with a single NXT microcontroller, more than 500 plastic pieces for building robots, 4 sensors (covered in Chapter 12), and 3 motors. The Education Kit comes with fewer plastic building pieces (about

400), but increases the number of sensors to 5 and provides a rechargeable battery pack as well as the single NXT microcontroller and 3 motors. View a more detailed side-by-side comparison of the two kits by visiting *http://legoengineering.com/content/view/42/90/* and clicking the **Hardware Comparison** and **Software Comparison** links. Keep in mind, however, that although the rules of FLL allow the teams to build the robots using only LEGO components, students can choose to use either the Retail or Education Kit. Most schools tend to purchase the Education Kit, though, so you'll typically find more teams using that kit.

What About the FLL Competition Mat and Models?

The mat and models come with the FLL kit sold by FIRST. This package contains the parts necessary to build the mission models used in the Robot Game portion of the competition as well as the vinyl field mat for the Robot Game. This FLL kit also consists of standard LEGO pieces and is only available from FIRST through its partnership with LEGO. Figure 1-6 shows an example of the mat and the mission models.

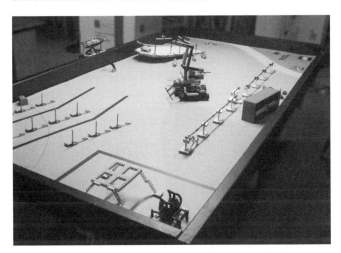

Figure 1-6: The competition mat has many missions to attempt, all built using LEGO pieces.

Teams love to receive the mat and the big bags of LEGO parts from FIRST and build the mission models. Since FIRST doesn't usually release the rules of the competition until after most teams receive their competition kits, teams often try to discern the purpose of the models and how the models might fit into the competition. Sometimes it's easy to figure out, and sometimes it's not.

Conclusion

The FLL Competition has been very successful in its endeavors. Since 1999, FLL participation has grown from around 9,000 students from the United States and Canada to over 100,000 students from 38 countries in 2007 (Figure 1-7).

Figure 1-7: Teams from around the world compete in FLL.

If you've never attended an FLL competition, you're in for a surprise. First, the noise level in the competition areas can be overwhelming. Kids and adults cheer their teams on as music plays, and commentators on loudspeakers detail each competition round. The excitement is contagious.

As you watch the students head off to the technical judging and presentation portions of the competition, remember that one of FLL's main missions is to help students to improve their speaking and writing skills. Self-esteem and confidence increase as students encourage and inspire one another to stand up, speak out, and show that they "know their stuff."

It comes as no surprise, then, that FLL has so much success attracting schools, parents, teachers, and students. The competitions are fun and exciting, and participants walk away with some truly astounding skills that will stay with them for life: research, problem solving, design, and public speaking combined with engineering and creativity.

2

HOW FLL WORKS

FLL is a unique competition. Unlike science fairs or robotics competitions, it consists of a wide range of team challenges that require individuals to combine many different skills to excel.

A new team may find the many aspects of FLL a little confusing at first. If this describes you, we'd like to help! In this chapter, we'll explain each category of the competition and how it works.

Although FLL is sometimes thought of as a robotics competition—like its counterparts FIRST Tech Challenge (FTC) and FIRST Robotics Competition (FRC)—robotics actually make up only half of the challenge. FLL has four different categories:

- Robot Game
- Project
- Robot Design
- Teamwork

These categories are combined into exciting competitions called *tournaments*. The following sections give an overview of tournaments and then describe the specifics of each category.

Tournaments

Tournaments are the climax of an FLL season. At a tournament, teams deliver their Project presentations, attend Robot Design and Teamwork interviews, and compete in the Robot Game.

Regions throughout the world host championship tournaments. The winners of these tournaments move on to the next level; qualifying tournament winners move on to the state tournament and so on. For example, in the United States, winners of state championships compete in national or international invitational tournaments.

The World Festival is a huge international invitational tournament composed of some of the best teams from around the world. Although it isn't the culmination of all championships, it is one of the most celebrated and prestigious tournaments. Learn more about upcoming FLL tournaments in your area on FLL's website, *http://www.firstlegoleague.org/*.

Robot Game

The Robot Game is one of the robotic categories of FLL, and it is one of the most well-known and exciting parts of the competition. For this challenge, teams have three to five months (the actual time depends on the region) to build and program a LEGO MINDSTORMS robot to autonomously accomplish several missions on a 4-by-8-foot playing field in two-and-a-half minutes. *Autonomous* means that the teams may not remotely control the robots. The robots compete for points by accomplishing as many missions as possible during a match.

NOTE *The two-and-a-half-minute time frame during which robots attempt their missions is called a* match. *A* round *consists of each team at a tournament completing one match.*

At the beginning of each new FLL season, teams receive a kit that includes a field mat, LEGO pieces, instructional materials, and other supplies to get started. Let's review these supplies, starting with the field mat.

Field Mat

The field mat is a flat, flexible, 4-by-8-foot plastic mat that makes up the area on which the robots compete. Graphics on the mat depict a virtual environment related to the competition theme. For example, the mat for the 2005–2006 Ocean Odyssey season pictured an ocean environment. Figure 2-1 shows the field mat used during the 2007–2008 Power Puzzle season.

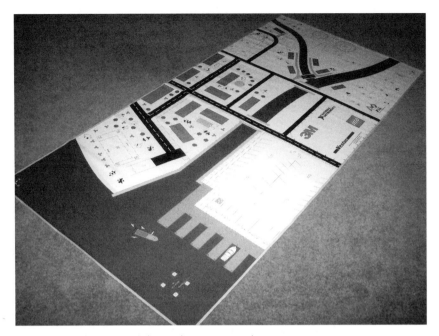

Figure 2-1: Field mat used during the Power Puzzle season

The mat should be set on a hard, smooth, and level surface, surrounded by 4-inch-high black borders (usually made out of 2-by-4-inch pieces of wood). At tournaments, the mats are usually set on custom-made tables, but teams can use other surfaces and borders for practice.

NOTE   *It's fairly easy to build an official table. Find instructions for doing so on the "Field Setup" page of the FLL challenge website. We suggest that you use an official table to practice to make your preparation as realistic as possible.*

Although most graphics on the field mat are decorative, some represent *scoring areas* where certain objects (or even the robot) must be delivered to or taken away from to score points. Other graphics are represented for strategic reasons. For example, black lines may be part of the virtual environment (such as roads), but a robot's Light Sensor may also use them for navigation. Look for features on the mat that might help your robot navigate.

Field mats also include an area called *Base*, which is the robot's starting point. The Base is a three-dimensional area 16 inches (40 cm) high, surrounded by graphics and one side of the field. Figure 2-2 shows the Base on the Power Puzzle mat.

At the start of each round, the team's robot must be completely inside the Base area, which restricts the size of the robot and ensures that all robots start each round in the same place. Also, if the robot must be rescued (as discussed in "The Match" on page 16), it must be completely inside the Base before it can autonomously start again.

Figure 2-2: The Base for the Power Puzzle season, bordered by three road graphics and one side of the table

Mission Models

Mission models are the LEGO constructions that are involved in all missions. For example, a mission might require robots to push a lever on a building made out of LEGO pieces, which is a mission model.

The Field Setup Kit contains a bunch of LEGO pieces that make up the mission models, as well as a CD with instructions on how to build them. For example, Figure 2-3 shows the mission model components for the Power Puzzle season.

Building the mission models is a great activity for a first team meeting as a warm-up or introduction. The building itself can take a few hours, so it might be helpful to print out the instructions so multiple people can build at the same time. On the other hand, if the team wants to start on the robot right away, one member may want to build the models during spare time.

Once you build the mission models, attach them to the field mat with Dual Lock—a fastening material similar to Velcro—which comes with your Field Setup Kit. The field mat shows where to place the mission models and the Dual Lock.

NOTE *If you're not sure how to use Dual Lock, FLL has instructions on how to use it, along with other useful field setup instructions, on its website.*

Figure 2-3: Mission model LEGO pieces for the Power Puzzle season

Mission models create a variety of challenges for robots, which usually have to perform the following four types of actions with the models:

- Transfer the model
- Activate the model
- Deliver an object to the model
- Remove an object from the model

Transferring a Model

This action involves moving a mission model from one location on the mat to another. When a mission involves this action, the model that will be moved is not attached to the mat with Dual Lock—it simply rests in a specified position. For example, in the Power Puzzle season, the Hydro-Dam had to be moved on the mat from Base to a river so that the model touched both banks of the river without touching any of the houses nearby.

Activating a Model

Activating a model involves doing something to make it react in a specific way. For example, in the 2006–2007 Nano Quest season, the Self-Assembly mission required the robot to flick a lever on the mission model to start a chain reaction. Models that require an activation action are usually attached to the mat with Dual Lock so the robots can activate them without moving the entire model.

Delivering an Object to a Model

Sometimes a mission model includes an object that isn't attached to the main model. Instead, this object usually starts in Base, and the robot has to deliver it to a specified place on the main model. In the Nano Quest season, for example, one mission required robots to deliver an object from Base to a mission model and drop it onto a lever that then caused a wheel on the model to spin.

Removing an Object from a Model

Some missions require robots to remove one or more objects from a mission model. For example, the Oil Drilling mission of the Power Puzzle season required robots to move three oil barrels from an oil platform and place them safely ashore. Points were deducted from a team's score if the barrels touched the water (ocean).

The Match

At a tournament, tables are set up with two field mats next to each other and their four-inch-high borders connected. The mission models are placed on the mats.

Two teams compete on a table—one on each field mat. Although they are next to each other, the teams only interact on one mission model that is on the border between their field mats. Sometimes, the two robots compete to accomplish the mission first (with the winner scoring more points). Other times, the teams work together to accomplish the mission, with both teams scoring the same number of points if they succeed.

Each team may have two members act as *drivers* of the team's robot. The drivers set up the robot in Base before the match, run the programs, and manage the robot during the match. Drivers can change places with other team members, but only two drivers can be at the table at once.

When the drivers are ready, the announcer counts down, "3 . . . 2 . . . 1 . . . LEGO!" and the teams start their robots. The robots have exactly two-and-a-half minutes to accomplish as many missions as they can.

MULTIPLE RUNS

Teams commonly program their robots to perform a mission or two and then return to Base, where the drivers can set it up for more missions. This strategy proves very helpful in competition. First, it enables the team to continually realign the robot so navigation errors don't accumulate throughout the match. Second, it allows the team to use simpler attachments that need to work for only one or two challenges (as discussed in Chapters 10 and 11). And third, it enables drivers to modify their strategy during a match based on the robot's performance; for example, retrying a failed mission or skipping a low-point mission to make time for one with a higher point value.

If a robot gets stuck, the drivers can rescue it by picking it up and returning it to Base, though they risk losing points. They may grab their robot without penalty if any part of it is already touching Base—whether to modify it, reposition it, run another program, and so on.

The winner of the Robot Game can be determined in different ways. Sometimes teams compete in three or more competition rounds for the highest single or combined score. Other times, the top scorers move into elimination rounds in which teams compete in pairs and the winners of each pair move through successive rounds until one team wins. It's very exciting to watch the Robot Game—and even more exciting to compete in it.

The Project

The Project is one of the two nonrobotic categories of the FLL competition and has a bit of a science-fair style. Teams research and solve a real-world problem based on the challenge theme and present their research and solutions to their community. *Community* refers to local organizations, governments, companies—anything. The teams are also encouraged to impact their community in their research area (discussed in the "Presentation" section) and will typically be asked about this during a judging session.

At the beginning of the FLL season, the details and rules of the Project are posted on FLL's website at *http://www.firstlegoleague.org/* (click the link for your country, go to the current year's challenge page, and click **The Project**). The rules describe the Project's theme, general topic, and any special activities the teams have to do as part of the Project.

Research

Once the Project rules are posted, each team needs to choose a topic related to the theme. For example, in the Nano Quest season, in which the general theme was *nanotechnology* (the manipulation of atoms and molecules), a team might have chosen "Nanotechnology in Medicine" as its topic.

Once the topic is chosen, research it. For many Projects, you need to suggest a solution to a problem. For example, in the Power Puzzle season, the Project required teams to come up with ways to make a building more energy efficient.

You can perform your research anywhere, but keep in mind that judges like to see original research, such as personal interviews with scientists or information discovered during the team's experiments. Be sure to take good notes because you'll include your findings in a presentation that you give to members of your community and, ultimately, the FLL judges (learn more about the research portion of the competition in Chapter 15).

Presentation

Once your team has enough information, create a presentation for the judges at your tournament(s). The only requirement is that the presentation is shorter than five minutes (including setup), but otherwise, teams can give just about any kind of presentation they want. Possible presentation styles include PowerPoint presentations, plays, videos, speeches—even operas!

Whatever your presentation, include as many team members as possible to demonstrate good teamwork and to show that the entire team contributed to the Project. The number of participating team members will probably affect your score. If many or all of the members participate, your score will most likely be higher than if only a few participate.

During the presentation, identify the real-world problem you researched and your proposed solution, and discuss the research you performed. Make sure the team members know the material well and can clearly and smoothly present their pieces without reading from notes.

The judges will then ask the team members questions related to the presentation. Among other things, they look for evidence that the members did all the work and have a good understanding of the information related to their presentation.

Sharing Your Project with the Community

In addition to presenting your Project to the judges, remember that part of the Project includes presenting your research and solutions to the community. One of the goals of FIRST is to get students interested in science and technology, so judges like to see evidence that a team has reached out to the community—the more, the better. Presenting scientific and technological results is important to scientists or engineers, and this aspect of the challenge gives team members valuable real-world experience in public speaking.

You can share your Project in several different ways. One common and effective way is to give your presentation for a group, such as a school, club, or even a senior living facility. Many seniors love seeing students learning about breakthroughs and accomplishments in science and technology that weren't around when they were young.

You might also consider adding some extras to your presentations. For example, give a short talk about FIRST and FLL, or offer more detail than you are able to give in the tournament (where presentations are limited to five minutes). You could even demonstrate your robot to make things more exciting.

Another kind of community outreach involves working with government or private organizations to impact your community in the area of your research. For example, if a team picked "Nanotechnology in Medicine" for its Project in the Nano Quest season, they could have talked to an area hospital about implementing some of the technology they researched. Even if the organization doesn't use your suggestions, the simple fact that you worked with them and tried to make an impact on the community is a nice addition to your Project. It will help your score, and best of all, it is a tremendous learning opportunity for the team.

Robot Design

In the Robot Design category of the competition, judges give a subjective score of the robot's design.

The judges will ask about the design of your robot in an interview called the *technical interview* (or *Robot Design interview*). For example, they'll probably ask how your robot works, how you built and programmed it, and how you overcame obstacles with its design. Many times they have a table set up with a field mat so you can show your robot in action. The judges look for well-designed robots that can accomplish missions in consistent, clever, and/or unique ways. They also look for evidence that the team members did all of the work on the robot (without the help of coaches or mentors) and evaluate the ways the team approached their particular design challenges.

Teamwork

The Teamwork category focuses on team dynamics—how team members work together. Your score for this category is partly based on how well you perform in an interview with the tournament judges. In evaluating teamwork, the judges consider the following categories, taken directly from the FLL rubrics (each is explained below):

- Team roles and responsibilities
- Gracious professionalism
- Problem solving
- Enthusiasm and member participation
- Understanding of FLL values

Judges have multiple ways of determining how well teams do in each category. For example, in the Teamwork interview, they may give the team a challenge, then make an evaluation based on how the team tackles that challenge. They also watch teams throughout the tournament and sometimes interview a team at their pit area (the team's home base).

Team Roles and Responsibilities

When evaluating the team in the roles and responsibilities category, the judges like to see teams that assign specific roles and distribute work among the members. For example, is there a Lead Programmer or Battery Manager? The judges also like to see members covering for each other as necessary. For example, if one member is sick, does another cover for that person? Chapter 6 discusses ways to determine team roles.

Gracious Professionalism

Gracious professionalism is one of FLL's core values. The *FLL Coaches' Handbook* describes gracious professionalism as follows:

> Gracious attitudes and behaviors are "win-win."
>
> Gracious folks respect others and let that respect show in their actions.
>
> Gracious professionals make a valued contribution in a manner pleasing to others and to themselves as they possess special knowledge and are trusted by society to use that knowledge responsibly.

Put simply, *gracious professionalism* means acting graciously and respectfully to teammates, other teams, and visitors to the competition. For example, if some of the teammates disagree about what to include in the Project presentation, the judges will expect the teammates to listen to one another and professionally resolve the disagreement. Gracious professionalism also applies to interactions with other teams (you might also call this *good sportsmanship*). For example, although this is a competition, the winning team should not attempt to put down the losing teams. By the same token, if one team loses a piece at the tournament and another team loans an extra to them, judges viewing the action will most likely increase the loaning team's score.

Problem Solving/Team Dynamics

The Teamwork judges are also interested in how teams overcome obstacles or solve problems they encounter. For instance, if a robot starts acting strangely the day before the tournament, the judges will probably be very interested in hearing how the team tackled that challenge. They look at how the members work together as a team, regardless of whether or not they ultimately solve the problem. The judges like to see evidence that all the members helped to solve the problem, working together and respecting one another's ideas.

Enthusiasm/Participation

During the interview, the judges look at how teams answer questions to see how enthusiastic and confident the members are. They also consider how many team members participate in answering questions; the more, the better, of course.

Understanding of FLL Values

Finally, judges consider how well team members understand FLL values and how much they learn from the FLL experience. For example, they listen for a demonstrated interest in science and technology on the part of the team members. They also want to hear that the team has learned useful things from participating in FLL, whether they learned how to build and program robots, how to present to large groups, or anything else.

Team Challenge

As mentioned earlier, sometimes the Teamwork category includes a surprise challenge that the team first learns about only in the Teamwork interview. These challenges differ for each tournament and can include just about anything. For example, one challenge in a tournament during the Nano Quest season required teams to build the tallest possible structure out of (uncooked) spaghetti noodles and mini-marshmallows. Teams have a limited amount of time to attempt the challenge. As they work on the challenge, judges look at how the members interact, how many participate, and how they perform in other aspects of the Teamwork category.

Awards

What's a competition without trophies? FLL tournaments usually give out several awards. They are often made out of LEGO bricks, like the trophy shown in Figure 2-4. Most tournaments give out at least the following five main awards:

- Robot Performance Award
- Robot Design Award
- Project Award
- Teamwork Award
- Champion's Award

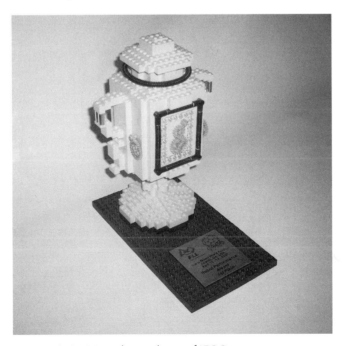

Figure 2-4: An FLL trophy, made out of LEGO pieces

The first four awards are given to the teams who do the best in the respective categories of the competition. The Champion's Award is considered the highest award in FLL, and it usually determines which team moves on to the next championship tournament. It is based on all four categories of the competition, and each category is weighed equally.

NOTE   *Teams can only win one of the Championship Tournament Awards, unless the winner of the Robot Performance Award, for example, also wins the Champion's Award. Each team may also win awards in only one championship series.*

Many tournaments give other awards as well. For example, some reward outstanding coaches or mentors. Others, such as the Rising Star Award, are awarded to teams that give an exceptional performance in some area even though they're not a winning team.

AUTHOR SNAPSHOT

Many of the suggestions in this book are drawn from Jonathan Daudelin's firsthand experience participating in FLL as the leading member of Team 1221: Built On The Rock during the 2006–2007 season. His team's robot achieved two perfect scores at the state tournament and won the Champion's Award. This award earned them an invitation to the World Festival, where Team 1221 and one other team made history by becoming only the second and third teams to achieve perfect scores in all three Robot Game rounds. Besides sharing the First Place Robot Performance Award, his team also won the First Place Innovative Robot Award.

3

GUIDELINES AND RULES

Just like any other competition, FLL has a set of rules and guidelines that describe the challenges. They tell teams what to do and what to expect at tournaments and help ensure that everything is done fairly and consistently. Since these rules and guidelines are so important, you should understand them well.

FLL has different kinds of regulations for different categories. Some guidelines don't describe the requirements of a challenge, but rather list what the judges will look for in the interviews. There are formal lists of rules as well as simple descriptions, and you can find the guidelines and rules in various places. This chapter lists each type followed by a discussion:

- Mission Descriptions
- Robot Game Rules
- Q&A and the Forum
- Project, Teamwork, and Robot Design Guidelines
- Rubrics

Keep in mind that the FLL competition changes its theme every year, so it is always possible that the rules may change too. The big picture you should take from this chapter is that it's very important for the teams and their members to familiarize themselves with the rules and organization of the current season's challenges.

Mission Descriptions

The Setup Kit you receive at the beginning of each FLL season provides the supplies and instructions for building the Robot Game's mission models; however, it doesn't reveal what the robot has to do with them. This detailed information is released later on the FLL website (*http://www.firstlegoleague .org/*) in the Challenge section. The *mission descriptions* state what your robot has to do for each mission and how many points the team receives for completing that mission.

Analyzing Mission Descriptions

Take a look at the following mission description for the Hydro-Dam mission from the Power Puzzle season.

> Mission: Place the Dam so it is TOUCHING both banks of any river section east of Base for **25** points. The Dam must be upright. When the match is over, the referee places (or projects) the Flood upstream of the Dam and there is a single maximum **10**-point deduction if any houses are being TOUCHED by the Dam or Flood. The Dam is never considered a stray object.

Notice several things about this description: First, it explains what to do for the mission. The robot has to place the Dam so that it touches both banks of a river section that is east of Base. The Dam has to be upright, and the Flood mission model (which resembles an area of water) can't touch any houses when placed in front of it. The description also lists the point value of the mission. Missions are designed with varying levels of difficulty, and therefore, different amounts of points are rewarded when missions are accomplished. In this case, the team gets 25 points if the Dam touches both banks of the right river section and is upright. However, you lose 10 points if the Flood touches any houses when it's placed in front of the Dam.

Why are *touching* and *touched* underlined and in all capital letters in the description? Did the writers think they had to yell for people to understand them? Actually, those are special words with specific meanings in FLL, and they are defined in the rules, which we'll talk about later. For now, just understand that when those words appear like that, they have a particular meaning.

Another thing to notice is how important it is to read the description carefully. Mission descriptions are usually very specific so that teams know exactly what to do. It's important to understand the whole description, so a team doesn't leave out one of the requirements for a mission and lose points for it. For example, if a team quickly reads through this description and misses the part about the Dam needing to be upright, they might put it on its side and consequently lose the points for that mission.

At the beginning of the year, after the mission models are built and set up on the field mat, it's a good idea for the whole team to review the mission descriptions. Make sure you know exactly what to do for each mission to avoid doing something wrong and losing points. It's also a good idea to print out a copy of the mission descriptions to take to tournaments, just in case you, another team, or a referee has a question about the rules.

Strategizing with Mission Descriptions

Once you read through the mission descriptions and understand the missions' objectives, make a plan for tackling them. This plan should include which missions you will attempt, the order you will attempt them in, the amount of time you want to spend on each mission, and so on.

Since the Robot Game is designed so that obtaining a perfect score (successfully accomplishing all the missions) is very difficult, your team might not attempt all the missions. If this is the case, choose to attempt the best missions—based on difficulty, point value, and time needed. To make this decision, go through the mission descriptions as a team. Look at the point value of each mission, and estimate the difficulty and time needed for each one. Then weigh the advantages and disadvantages, and pick the best overall missions. For example, one mission might award 50 points but be difficult and time consuming. Another mission, on the other hand, might award only 35 points but be quite easy and close to Base. Look at how much harder and longer the first mission is to see if it's worth the extra points.

NOTE *In many FLL seasons, 400 is the maximum number of points a team can achieve in the Robot Game.*

Once you pick your missions, you may find it helpful to plan the order in which the robot will attempt them. Some missions might work better if the team attempts them at the end of a match, while you need to accomplish others right away. For example, if the two-team mission (the mission between the two adjacent fields on each table) is a competition and you need to accomplish it before the other team, you will probably want to attempt it as soon as the match begins. On the other hand, it might take a long time to get to one mission model, so you might save it for last so the robot doesn't have to return to Base or go on another mission. Robots can be anywhere on the field at the end of a match.

It's also a good idea to make a rough estimate of how long each mission will take the robot to attempt. Remember, you only have two-and-a-half minutes for each match. If your estimates show the robot taking significantly more or less time to attempt all the selected missions, consider adjusting the number of missions so the total time is closer to two-and-a-half minutes. Having time estimates is also helpful when you build and program a solution to a mission.

Robot Game Rules

The Robot Game rules section could be described as the *nitty-gritty* of the Robot Game, and it mainly defines the formal terms and rules. Find the rules by going to the website for this year's challenge and clicking **Rules**. Carefully read through these and understand all of them, as they describe many things you may and may not do, as well as what you're required to do. Just like the mission descriptions, it's a good idea to print the rules and take them with you to tournaments.

One important rule described in this section concerns the materials that teams are allowed to use. For example, the *Materials* rule for the Power Puzzle season said the following:

> **7. Materials** This rule is not just about the robot . . . This rule controls everything you bring from the pit area to the competition area including the robot, all attachments, and all strategic objects when viewed all at once as a package. All these objects must be made entirely of LEGO elements in original factory condition (except LEGO string and tubing may be cut to length), and must conform to the following quantity limits on electrical parts, no matter what you intend to use or connect or attach to the robot at any one time:

For RCX users:	For NXT users:
RCX controller (1)	NXT controller (1)
motors (3)	motors (3)
touch sensors (2)	touch sensors (2)
light sensors (2)	light sensors (2)
lamp (1)	lamp (1)
rotation sensors (3)	rotation sensors (3 minus
3rd touch OR light	the number of NXT
sensor (1)	motors present)
	ultrasonic sensor (1)

> LEGO wires and converter cables are allowed as needed. Spare/alternate electrical parts are allowed in the pit area. Objects functioning as remote controls are not allowed anywhere. There are no restrictions on the quantity or source of non-electric LEGO pieces. Wind-up/pull-back "motors" are allowed, and do not count as motors. Pneumatics are allowed. Marker may be used for owner identification in hidden areas only. Paint, tape, glue, oil, etc., are not allowed. Stickers are not allowed except LEGO stickers applied per LEGO instructions.

This rule basically says you can use only and all LEGO pieces, except there is a limit on how many sensors, motors, and intelligent bricks (RCX or NXT Brick) you can use. These limitations are very important. You can be disqualified if you exceed the limits or use non-LEGO parts.

Other important rules describe what team members can and cannot do during a match. These rules include *Participation, Housekeeping, Preparation Mode, Starting Position, Starting Procedure, Transition Mode, Muscle Action,* and *After the Match.*

The *After the Match* rule is particularly important. Once a match ends, referees look over the table and determine your score. The team members look it over at the same time and make sure they agree with the awarded score. Referees can and do make mistakes sometimes, so it's important to watch how they score you. No sense losing points because of a mistake in scoring!

CHECK YOUR SCORE

I've been a referee numerous times at FLL competitions, and I'll admit that I make mistakes. I think it's important for team members to check my scoring sheet and verify that I gave them the proper points for successful missions. Referees should never discourage teams from double-checking their calculations, so don't be afraid to ask.

—James Floyd Kelly

Q&A and the Forum

Although the guidelines and rules explain most things about the competition, teams often have a question that isn't clearly answered or a question about the rules themselves. On the FLL website, there is a section called *Q&A* which teams use to resolve these kinds of questions. You can find the Q&A on the Challenge page of the FLL website. The Q&A is a list of past questions about the competition, along with FLL's official answers. To submit a question, contact FIRST through email or phone; contact information can be found on the Q&A site.

Even if you don't have a question, you should regularly check the Q&A. All of the answers there become part of the official rules and take precedence over both mission descriptions and rules. Also, you might learn something important that you didn't know before or hadn't thought of from somebody else's question. Sometimes you can even learn useful strategies from the Q&A, since teams will sometimes ask if a certain strategy is permitted. Print out the Q&A, and take it to tournaments.

A great place to get unofficial answers to questions, as well as help with administrative stuff such as fundraising or registering for tournaments, is the FLL International Forum. Many people, some of whom are very experienced with FLL, post on this forum, and they can be of great help if you have a problem. You can find the forum on FLL's website. You must register to get started, and you'll need the forum access/setup code from your *FLL Coaches' Handbook.* This Handbook should come with your Field Setup Kit.

Guidelines for Other Categories

The guidelines for the other categories are not as detailed or quantitative as the rules for the Robot Game. These challenges can be carried out and judged with greater degrees of freedom.

The Project

You can find the guidelines for the Project on the Challenge website. This page describes everything you must do for this category. Unlike the Robot Game, in which the guidelines and rules are split up over several different pages, all the requirements for the Project are posted on this page. You can also find resources of helpful information related to the Project on the Challenge website.

Robot Design and Teamwork

FLL doesn't have specific websites that describe what to do for Robot Design and Teamwork, because the only thing that is required is to go to the interviews. However, there are criteria describing how teams are judged in these categories, which you can think of as good suggestions for improving your score! Find some of these guidelines by clicking **Awards** on the FLL website, including the following about the Robot Design award:

Robot Design Award

Judges look for teams whose work stands out for innovation and/or dependability. To assess innovation, the judges watch the robots work and look for things that make them say "Wow!" They interview team members to reveal the less obvious unique and inventive ideas. To assess dependability, the judges interview the teams to learn what solid principles and best practices they used to reduce variability and errors, with preference to robots that best "back it up" throughout the matches.

Tournaments may choose to break the Robot Design Award into two separate awards for programming and design, or those involved may choose to give three separate awards for programming, innovative design, and dependable design.

Farther down the page, it talks about the Teamwork Award:

Teamwork Award

Teamwork is critical to succeed in FIRST LEGO League and is the key ingredient in any team effort. FLL presents this award to the team that best demonstrates extraordinary enthusiasm, an exceptional partnership, and the practice of FLL values. The team receiving this trophy demonstrates the following attributes to the judges:

- Confidence, energy, and enthusiasm
- Group problem-solving skills
- Understanding of and respect for others
- Positive team interaction and group dynamics
- Demonstrated interest in science and/or technology
- Ability of team members to fill each other's roles when necessary

The above descriptions are basically summaries of the rubrics, which we cover next.

Rubrics

The rubrics are detailed lists of criteria the judges will use during the Project, technical, and Teamwork interviews. You can find the rubrics in the *Coaches' Handbook.*

The rubrics list many different aspects of each category that the teams are judged on, as well as suggested levels of performance for each aspect. Remember, the judges are individuals, and their judgments are subjective. They will use some or all of these rubrics in evaluating teams in each category, so having them is a great advantage. It's helpful to read through all the sections in the rubrics to get an idea of what the judges want to see.

4

FINDING EQUIPMENT, MENTORS, AND FUNDING

As your team progresses through the FLL season, you'll discover that certain resources can be extremely beneficial to your success. We group these into three categories: equipment, funding, and team mentors. Some of these resources are optional, while others are absolutely critical.

Finding funds and gathering resources isn't always the most glamorous aspect of participating in FLL, but some indispensable items, such as laptops, pens, paper, and plenty of LEGO bricks and TECHNIC pieces, cost money, so raising funds and finding equipment is definitely an important activity. In addition to finding the needed supplies, a team also needs to find a way to pay for them. And once a team has the necessary equipment, sometimes it lacks the skills or knowledge to get started. This is when team mentors can come in handy. Read on for advice on where to find these resources.

Equipment and Supplies

At a minimum, an FLL team can compete with a single NXT robotics kit, a desktop or laptop computer, some batteries, and a few cardboard boxes. FLL has done a great job of creating a competition that doesn't require a large initial purchase yet still provides a rewarding experience for new teams without many materials. These supplies may come from donations, fund-raising, or schools (which often provide computers and work space).

The following is a suggested list of equipment and supplies that an FLL team might find beneficial during competition. As you'll see, some items are required and some are optional. While the list certainly varies from team to team, this should give you an idea of the items to start collecting or raising money to purchase:

Required

- FIRST LEGO League Competition Package (mat and models)
- LEGO MINDSTORMS NXT Robotics Kit (Retail or Education version)
- Desktop or laptop computer
- AA batteries or a rechargeable battery pack

Optional

- Writable media for backing up data—CD-RW/DVD-RW or Flash drive
- Pens, pencils, poster board, paper, and so on
- Containers for storage and transportation
- LEGO MINDSTORMS NXT Education Resource Set (a parts kit not to be confused with the Education Kit)
- Camera and/or digital video recorder
- Extra LEGO bricks and TECHNIC pieces or kits

> **MORE POSSIBILITIES WITH THE EDUCATION RESOURCE SET**
>
> The *Education Resource Set* is a collection of TECHNIC beams, wheels, pins, and other components. The Resource Set does not come with motors, sensors, or the NXT Brick, however. Its purpose is to provide more parts (over 600, in fact) and can enable a team to use a larger variety of possible robot designs.

Before purchasing anything, create a list of the items you need, and determine which you may obtain from team members (or parents). Whether you purchase items or borrow them, treat all equipment and supplies with care; these items must last beyond just the FLL season.

> **FORM AN ENVIRONMENTALLY FRIENDLY TEAM**
>
> For the Project portion of the competition, consider demonstrating your team's "green" side. Recycle aluminum drink cans and wastepaper during team gatherings, document how your team uses rechargeable batteries and has its computers/laptops enter sleep mode when not in use, and find other ways to conserve resources. Do it not for extra points but because your team wants to do its part to help the environment.

Consider using a sheet like the one shown in Figure 4-1 to list the supplies your team needs and track the sources and locations of items.

NOTE *Download the Resource Tracking Sheet at* http://thenxtstep.com/book/downloads/.

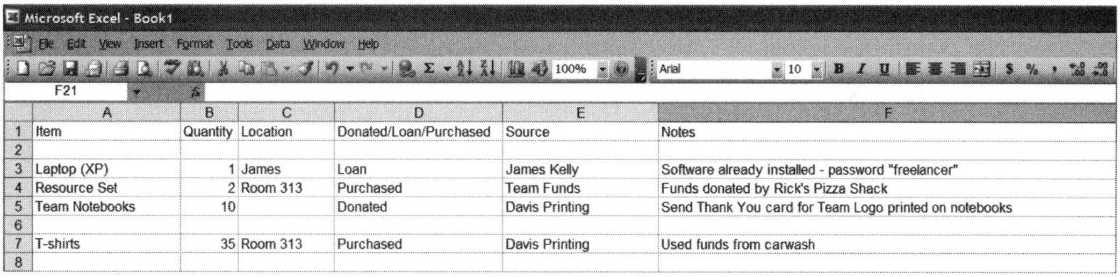

	A	B	C	D	E	F
1	Item	Quantity	Location	Donated/Loan/Purchased	Source	Notes
2						
3	Laptop (XP)	1	James	Loan	James Kelly	Software already installed - password "freelancer"
4	Resource Set	2	Room 313	Purchased	Team Funds	Funds donated by Rick's Pizza Shack
5	Team Notebooks	10		Donated	Davis Printing	Send Thank You card for Team Logo printed on notebooks
6						
7	T-shirts	35	Room 313	Purchased	Davis Printing	Used funds from carwash
8						

Figure 4-1: Keep a Resource Tracking Sheet to help organize your team's supplies.

A Resource Tracking Sheet can help a team track its resource usage and determine when to obtain more supplies. You can also use it to ensure that loaned items are returned at the end of the season (using the "Source" section) and that thank-you cards are sent to acknowledge donations and loaned items.

Funding

A team's funds can come from a variety of places—schools, government (local, state, or federal) grants, parents, businesses, fund-raising activities, and community donations. Some funds will be designated for particular items (such as the purchase of a computer or team shirts), so be certain to spend funds properly if the source is specific about usage.

Schools

Some schools donate funds to FLL teams, while others donate supplies, equipment, or meeting space. Most schools must account for all funds, so a team coach may need to track all expenditures.

School board meetings are a good place for a team to present information about FLL. Perhaps you can convince your school board to allocate funds to FLL by demonstrating some of the benefits of FLL, accompanied by an itemized list of needed equipment and supplies and a budget.

Schools also often allow teams to hold fund-raisers such as car washes on their grounds. This usually requires sponsorship by a school administrator or teacher.

Government

Funding by local, state, and federal government varies. You need to do some digging and make some phone calls to track down funds, like grants. For information on locating and getting grants, try *http://www.grantwrangler.com/* and *http://www.schoolgrants.org/*. Also check out *http://www.k12grants.org/ Grants/federal.htm* for federal grants (United States only) and *http://www .k12grants.org/Grants/state.htm* for state-based grants (click your state's name to begin). Be certain to check out the rest of the *http://www.k12grants.org/* website: It's full of other opportunities, such as foundations and special interest organizations that provide funding.

Parents

It's no surprise that a large portion of a team's resources often comes from the team members' parents. Whether it's a gift of money or equipment, parents don't seem to hold back when it comes to encouraging their children's interests.

Sometimes fund-raising is as easy as adding up the necessary money, equally dividing the sum among the team members, and asking each family to contribute its share. Financial situations differ from family to family though, so be sensitive; it's usually best for the coach to approach each family and determine the possible level of financial support.

Alternatively, consider dividing up the total funds required, and ask each team member to try to raise those funds by obtaining sponsors (see "Businesses" below). It's not difficult for parents and team members to email polite and professional letters (or mail handwritten letters) to various organizations to inform them about the FLL team and its fund-raising attempts. Start simple—email or send letters to the restaurants, grocery stores, banks, and other organizations with which your family does business.

Team members should also approach their parents' places of employment to inquire about sponsorship. Many employers are happy to provide funding or equipment to the activities of their employees' children.

Businesses

Team members should approach their parents' places of employment when trying to raise funds—but don't stop there. All businesses in your area are potential sources of funding. The chances of success are even greater when you approach businesses that your family frequents; if the employees know your name or always recognize you, that's an ideal business to approach about sponsorship.

How much should you ask for from each sponsor? Let's assume your team needs to raise $5,000. It's a large amount, and if you only ask 10 sponsors for funding, each will need to contribute $500. But consider what happens if you approach 100 sponsors; now each sponsor is only being asked for $50. Which amount do you believe will be easier to ask for and receive? And if you have a team of 10 members, this means that each member only needs to find 10 businesses willing to donate $50. That might not be too difficult.

If you're looking for a way to approach businesses about sponsoring your FLL team, Figure 4-2 shows a sample letter that you can email or send to area businesses. Be sure to add personal comments to remind the readers of your family's relationship with those businesses. (It's also a good idea to offer some sort of advertising in return for sponsorship. For example, many teams make shirts with a list of their sponsors printed on the back.) Sign the letter, possibly include your phone number, send it off, and wait for a response.

FIRST LEGO League

TEAM NXTSteppers

May 29, 2008

Rick's Pizza Shack
123 Main Street
Somewhere, OK 10101

Dear Mr. Rhodes,

My name is Decker Kelly, and my parents and I are huge fans of your pizza. I have a deal with my dad that every time I cut the grass, he'll order me a small Jamaica Pizza (my favorite).

This year, I'm participating in the FIRST LEGO League robotics competition as a member of the NXTSteppers. My team is currently raising funds to buy parts for our robot and team shirts, and I was wondering if Rick's Pizza Shack might be willing to sponsor my team with a donation of $50?

My coach's name is Erin Dennis, and her phone number is 404-555-1234 if you have any questions. When the season starts, I'll be sure to call and invite you to come to watch one of our competitions, and then maybe the team can have a pizza party when we're done.

Thank you!

Decker Kelly

Figure 4-2: Send a personal note to potential sponsors to raise funds.

NOTE *Download this sample letter at* http://thenxtstep.com/book/downloads/.

Remember to equally spread out the work of obtaining sponsors among the team members, and try to keep the sponsorship fee low for the best results.

Fund-Raisers

Our last suggestion for raising money and equipment is organized fund-raisers. FLL teams have used all sorts of fund-raising methods, including car washes, bake sales, and raffles.

The key to a successful fund-raiser is making sure that all team members are involved. Pick dates for fund-raising activities when all the team members (including parents, if possible) can attend. While the team members should do the actual work, parents can help with handling money and answering questions.

Watch for local events and festivals at which your team might be able to set up a booth to sell items such as cold drinks (on a hot day) or food items. Be sure to put out a donation jar; buyers often throw their change into it. And don't forget to have some information printed (brochures or posters) that provides details about FLL and your team's participation.

TEAM BUSINESS CARDS

Some teams print business cards with their team name, number, and logo on one side, and the FLL values on the flip side (phone number and email address are optional). Each team member carries a handful at all times to distribute to potential sponsors, friends, and family. Business cards are not only a great way to start a conversation about FLL, but they're also a professional method for team members to introduce themselves to sponsors, judges, other teams, and anyone unfamiliar with FLL. The online printing store at *http://www.48hourprint.com/* has great prices and quick turnaround, or you can contact a local print shop.

—JFK

Team Mentors

All teams begin with team members and a coach or multiple coaches. The skills that team members and coaches bring to the FLL competition aren't always related to building and programming robots, however, let alone the FLL theme for which the research will be focused. Sometimes a team needs help with FLL technical requirements or other skills such as graphic design, promotion, and fund-raising.

The team mentor can provide assistance in these areas and more. Mentors are individuals from whom teams can learn. The mentor's job is to offer guidance and training to the team members, not to do the actual work.

Some mentors provide teams with training in basic programming theory or mechanical design, allowing the members to get started building and programming robots. Other mentors provide teams with assistance in designing logos, shirts, and presentations.

Some mentors have a specific background in an area related to the FLL team and can answer questions or provide more details on a research topic. For example, during the Power Puzzle season, a team might have asked a mentor who installs solar panels to help with solar panel research.

Mentors sometimes have limited availability, and you can't expect them to attend every team gathering. Often a mentor may only be able to meet with the team on weekends or evenings. Most mentors are donating their time, so always remember to respect that time and to make your team available when your mentor is available.

Where can a team find good mentors? Start locally by investigating the following:

- Colleges or universities in your area
- Professional organizations
- Family members working in areas related to the FLL theme
- Businesses with services or products related to the FLL theme

Always explain the concepts of FLL to potential mentors. Be specific when requesting services or skills, and provide mentors with a list of meeting dates and times so they can plan their schedules.

A good team mentor not only provides skills and knowledge but also serves as a good role model and motivator to the team members. Team members should ask questions and learn more about the mentor's training and work. If you wisely use the time with a team mentor, the entire team will benefit.

Gathering Resources Early

The team may not have all of the necessary resources at the start of the FLL season, but because FLL releases the new theme well in advance, teams can often get a jump start by locating and gathering the resources early. Once the team begins to acquire equipment, supplies, funding, and mentors, it can then shift its focus to the Research and Robot Game portions of the competition.

5

STARTING OR BUILDING A TEAM

Assembling an FLL team can sometimes be as simple as pinning a signup sheet to a classroom wall or asking a group of friends to participate. But the reality of the situation is that there's often much more to organizing an FLL team than a simple show of hands or list of names. Sometimes a team isn't formed as quickly (or easily) as you might think. Recruiting and selecting team members can be time consuming. Sometimes more team members want to participate than there are team slots. Then there's the administrative issue of registering the team.

This chapter provides some suggestions not only for forming a new team, but also for selecting members and building an existing team. The methods discussed are certainly not the only ones you can use, but they are the result of the experience of several FLL coaches and teams that have struggled with these issues.

We start by discussing building a team from the ground up and then move into managing an existing team in which rookies may join veteran members. By the end of this chapter, you should have enough information to form a team and start the season.

Methods for Starting a New FLL Team

Although FLL teams can be created in many different ways, the following covers some of the more common ways.

School Programs

Many schools participate in FLL with students from a math, science, or other class. This is the most common start for an FLL team; schools frequently draw on their teachers to act as coaches, with parents providing assistance. Schools provide resources such as computer labs, robotics kits, and meeting spaces—all of which a team will require during the season.

After-School Programs

After-school programs can take on many forms. Sometimes a school provides a program, but often a nonprofit or other business entity supports an after-school program. Like schools, programs such as this can usually provide resources such as computers and robot kits for the teams to be productive.

Community and Church Groups

Check with your local community centers and churches to determine if they offer a team. Civic organizations and churches often enjoy sponsoring teams as a means of reaching out to the community. Sometimes funding is extremely limited, though, so teams often have to provide their own computers, laptops, and robot kits.

Homeschool Groups

Many homeschool groups and families participate in FLL for both the experience of competition and for school credit. Homeschooled students must not only make certain that they have access to resources the team will need but that they also fulfill any educational requirements (such as tracking the time spent on activities) as defined for homeschool programs.

Motivated Parents or Children

Sometimes a parent or child aware of FLL may be all the impetus needed, many times using one of the above means to assemble and support a team. Parents who wish to involve their children have even created small teams of two or three siblings.

Getting Started

Starting a new team from scratch can be easier than reorganizing an existing team. Existing teams frequently have a "this is how we do it" attitude, and new members may feel overwhelmed and a bit shy about expressing their opinions and suggestions for improvements. New teams do not have to be concerned with predefined rules or methodologies, which can allow the team some real freedom to explore new ways to work together. So let's build a team.

The following are the steps we follow when forming a new team:

1. Announce that a team is being formed.
2. Begin accepting applications with a well-defined deadline.
3. Hold your first team meeting (covered in more detail later in this chapter and in Chapter 6).

Announcing Formation of a Team

You can announce that you are forming an FLL team in many ways. Some common methods include the following:

- Announcement to parents in a school newsletter
- A community notice in a newspaper
- Phone calls to friends
- An after-school program mailer

No matter which method you use, be sure that it reaches enough people so that you have enough to field a team. FLL teams are limited to 10 students, and depending on your announcement method, too few or too many students may sign up (see "Selecting Final Team Members" on page 46 for help if you find you have more applications than team slots).

Either way, ensure that no students are left out of the application process because they weren't aware of the announcement. Also, depending on how much adult help the coach will have, it may make sense to limit the number of team members for manageability; you don't need to have 10 team members to have a successful season.

When soliciting participation, you may want to include a deadline for receiving and reviewing applications, but be careful about making exceptions. Applicants should feel that all the team members are participating in a fair selection process, and early acceptance of any student applications could bring charges of favoritism.

Finally, briefly explain how you will accept applications. For example, if you will require applicants to write a short essay explaining why they want to be on the team, mention this in the announcement to allow applicants to begin preparing. Likewise, if you will use a simple signup, share that information as well.

Figure 5-1 shows a sample email announcement, including relevant information that an applicant needs to satisfy the application process (download this sample announcement at *http://thenxtstep.com/book/downloads/*).

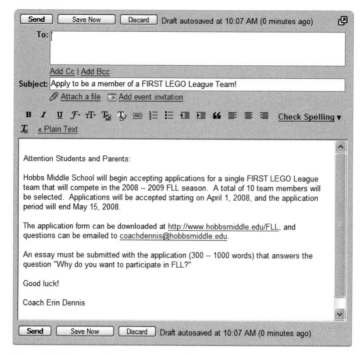

Figure 5-1: An email announcement for a new FLL team

Accepting Team Member Applications

After you announce the team and establish a date for accepting applications, it's time to start collecting applicants. Possible methods for accepting applicants include one or all of the following:

- Signup sheet
- Printed application
- Interview
- Essay

Signup Sheet

A *signup sheet* is simply a list of people who want to become team members. However, using a signup sheet alone can cause some problems. For example, will team members be selected based on the order in which they sign up? If so, what if a student is sick on the day you post the sheet?

Signup sheets also provide little information about the student's skills or commitment to the new team. If you use a signup sheet, use it only as a method to gather information about interest and the potential size of the team. Then allow for other factors, such as an essay and skills assessment (as discussed below), to help select applicants.

Figure 5-2 shows a sample signup sheet (download this signup sheet at *http://thenxtstep.com/book/downloads/*).

FIRST LEGO League Signup Sheet - Hobbs Middle School

Instructions: **Please add your name and grade to the list below. You may only add your own name to the list – do not write in any name other than your own. The signup period will end on May 15, 2008, and a list of selected team members will be posted on May 31, 2008.**

NAME   **GRADE**

1. _____

2. _____

3. _____

4. _____

5. _____

Figure 5-2: A signup sheet is a good start for collecting the names of interested students.

Printed Application

Requesting that prospective team members complete a printed application can help a coach whittle down a large applicant pool by providing insight into each applicant's commitment and skills. This application would request parent names, phone numbers, and email addresses, as well as an additional applicant requirement, such as a written essay (for example, "Why do you want to compete in FLL?") or a checklist of skills the applicant brings to the team. Figure 5-3 shows a sample application.

Interview

Interviewing applicants is a great way to find the most qualified applicants, but it's also extremely time consuming. For example, if the interviewer grants 10 minutes to each applicant, interviewing 30 applicants will take more than 5 hours. And this doesn't include the actual selection process in which you review your notes and pick the best candidates. If you decide to conduct interviews, do so only if time permits, and be certain to give every applicant the same amount of time and identical questions. Take good notes on each applicant's responses too.

One option in lieu of an interview is a questionnaire (which should supplement the team application, not replace it). Figure 5-4 shows a sample with a list of 10 questions. Applicants are limited to two to three sentences for each response to a question. The reviewer assigns a value between 1 and 10 based on the quality of each of the applicant's responses and then adds up the values to calculate that applicant's total questionnaire score. These scores can then be used as an additional factor in the decision-making process (download this questionnaire at *http://thenxtstep.com/book/downloads/*).

SKILLS ASSESSMENT

Jessica Mallard, a fifth-grade teacher at Wichita Collegiate School in Kansas, suggests including a "skill cloud," which is a list of words the applicant can circle to indicate skills and interests, on the application. See an example of a "skill cloud" in Figure 5-3.

> For students with no experience, I have them circle words in a word cloud that describe them best (logical, artistic, hands-on, mechanical, communicative, etc.) and place them based on whether the answers are more "programmer-like" or "builder-like." More experienced builders and programmers become Builder 2 and Programmer 2 in their respective groups. Less experienced students are Builder 1 and Programmer 1.

FLL TEAM MEMBER APPLICATION

PERSONAL INFORMATION

Student Name

Parent/Guardian Name

Street Address

Email Address

Telephone Number   Birth Date

Gender
☐ Male   ☐ Female   Email Address:

EXPERIENCE

Have you previously participated in an FLL competition?   Yes ☐   No ☐

If you answered Yes, please describe your participation, including any duties you performed:

ESSAY

Please write an essay (300-1000 words) that answers one of the following questions (pick one question ONLY):

1. Why do you want to participate in FLL?
2. What special skills or interests do you possess that will benefit an FLL team?

SKILL CLOUD

Circle three words that best describe your interests and/or personality

Artistic   Organized   Hardware

Record Keeping   Software   Design

Enjoy Research   Big Groups

Novice Programming   Public Speaking   Novice Building

Small Groups   Advanced Programming

Advanced Building   Inquisitive   Take Things Apart

Computers   Creative

DEADLINE: Your application and essay MUST BE received by Friday, May 15, 2008.

Please mail or FAX this completed application to:
Erin Dennis, Coach   FAX: 541-779-2018
Hobbs Middle School   Email: coachdennis@hobbsmiddle.com
123 Main Street, Atlanta, GA 30303

Figure 5-3: Use an application to let applicants describe their skills and experience.

FLL APPLICANT QUESTIONNAIRE

(PROVIDE A MAXIMUM OF 2 TO 3 SENTENCES PER QUESTION.)
Question 1: What previous experience do you have working with a team (any type)?
Question 2: What do you think might be the most enjoyable part of being a member of an FLL team?
Question 3: What is the most important personal skill or interest you believe you could offer a team?
Question 4: Describe how you best learn new subjects.
Question 5: Describe an experience on a team where you learned a new skill or gained knowledge on a topic.
Question 6: On an FLL team, for what position or job do you feel you are most qualified or interested in?
Question 7: What subjects do you enjoy most (can be technical or non-technical)?

Figure 5-4: A questionnaire can be useful during the selection process.

Essay

An essay can give you just as much insight into a student's excitement and commitment to an FLL team as an interview. An essay can provide more detail about an applicant than the short responses on a questionnaire, but you are typically limited to one or two essay topics. Some possible essay topics include the following:

- Why do you want to join an FLL team?
- What are your expectations as a potential FLL team member, and what skills or talents do you bring to the team?
- Describe a school project or personal project that you worked on; include a description of the types of work you performed and the results of the project.
- What are some of your favorite subjects, and how would you incorporate them into a position on an FLL team?

Prepare for the team member selection process by combining all the applicants' essays with their respective applications, interview notes, skills assessments, or questionnaires.

Selecting Final Team Members

If more than 10 participants apply for your team, making the final selection of team members can be extremely difficult and stressful. Every coach selects members differently, and there are no right or wrong methods for narrowing down the list of applicants. Here are some factors you might consider when making your decision.

Experience and/or Seniority

Consider giving preference to older students since they will have less time to participate in future FLL competitions. FLL team members must be between the ages of 9 and 14; during the selection process, you might take an older applicant's age into consideration since he will most likely not be able to participate the following season.

Unique Skills

If all but one of your applicants indicate they want to build robots, and the remaining applicant indicates she has experience with programming, this rare interest has already shown the applicant to be an asset to the new team.

Likewise, the research portion of the competition often gets less attention than the Robot Table work. If an applicant indicates a strong interest in reading, writing, and public speaking, he might make an excellent addition to the team as its Research Team Leader.

Gender Balance

All-girl and all-boy teams exist, but when the applicant pool is of roughly equal size and gender, consider a balance when making selections so one gender doesn't overshadow the other.

Displayed Interest

You will find that many applicants sign up to join the team because their friends applied or because they have a passing interest in robotics. While these are acceptable reasons for wanting to participate, it's important to select team members who demonstrate a sincere interest in the FLL experience or an aptitude for science, technology, math, and science.

Announcing Your Selections

The announcement of the final team members is an exciting time for those selected, but it's also a disappointing time for those not chosen. Anytime the applicant pool is greater than the number of team members that can be supported, there will be hurt feelings. What can you do about those who want to participate but don't fit on your team?

One way to meet the desires of those not selected is to form a second team (or even a third or fourth, as necessary) so that no student is denied participation in FLL. After all, FLL is meant to be inclusive, not exclusive. The more teams, the better.

Another option is to consider selecting two or three alternates who can participate in team meetings as observers and step in when a team member is absent or leaves the team (be certain to specify the order that the alternates would join the team to avoid later disagreements).

Whatever you do, remember that your new team has been created and that its members are excited and ready to begin. While "there's always next year" for those who were not selected, your new team is looking at this year. Be sure to keep their needs in mind.

Holding Your First Team Meeting

Congratulations! Your new team is formed. But the team can look forward to a lot of work, and it all starts with the first team meeting. Many coaches and students use the term "controlled chaos" to describe this first meeting; it is often a fast-paced gathering in which information overload can occur.

What should the team try to accomplish during the first team meeting? Below are some suggestions for the first gathering that can help a team organize and plan more efficiently (we cover some of these in more detail in Chapter 6).

Introduce Coach, Team Members, Mentors, and Parents

Make certain that you introduce all the participants. While the team members may know one another, this is often the first time for the coach, parents, and mentor(s) to meet. Facilitating these introductions can help when it comes to arranging travel, picking future meeting locations, and coordinating competition day activities.

Exchange Contact Information

The coach should obtain and make a list of contact details for all the team members, including the coach, mentors, and parents. Contact information should include mailing addresses, phone numbers, and email addresses. Parents and mentors may also want copies of this information. Consider passing around copies of a form for the participants to fill out; make photocopies or type up the information, and email it to all the participants.

Discuss Meeting Locations and Schedule

This is an ideal time to talk about the frequency of team meetings and to schedule future team gatherings. Consider using an Internet-based group calendar such as Google Calendar for sharing this information and competition dates (see Chapter 8 for more about using Google Calendar).

Assign Team Member Roles and Duties

Many activities need to be performed during an FLL season, and assigning team roles and duties will help manage the workload. Chapter 6 covers this topic in more detail.

Sign a Team Member Pledge

Some teams have a document, or a *code of conduct*, that spells out certain rules of behavior. If a coach wants to have the members sign this document, review it during the first meeting, and make certain that the students and parents both understand the team requirements and expectations of behavior. Documents like this can frequently come in handy when handling disagreements of opinion or disruptive teammates. Figure 5-5 shows a sample Team Member Pledge (download this document at *http://thenxtstep.com/book/downloads/*).

FLL TEAM MEMBER PLEDGE SHEET

As a member of FIRST LEGO League Team # _____ , I pledge to honor the following FLL Values:

Respect each other in the best spirit of teamwork

Behave with courtesy and compassion for others at all times

Honor the spirit of friendly competition

Act with integrity

Demonstrate gracious professionalism

Encourage others to adopt these values

I make a promise to my coach and fellow team members to behave appropriately at all team meetings and to respect my fellow team members' opinions, decisions, and work.

I understand that the team's success will depend upon the combined efforts of all team members, and I pledge to provide assistance when asked.

Above all, I pledge to be a good team member and to provide my fellow team members with my best efforts, energy, and attitude.

_____
Signature   Date

Figure 5-5: A Team Member Pledge can help remind participants of their team obligations.

Don't Forget the Food!

If the first meeting isn't held at a restaurant (pizza parties are a favorite first meeting place), consider asking one or more individuals to arrange for catering or snacks. Sometimes these first meetings can last longer than expected, so be sure that food and drinks are available.

FLL should be a fun, enriching, and educational experience, and the coach and team members should always strive to make every aspect of a new team's formation (and subsequent meetings) friendly and inviting to all members.

Adding New Members to an Existing Team

When adding new members to an existing team, many of the tasks involved are identical to those for creating a new team, but we'll shortly cover a few differences. Below is a modification to the steps listed earlier in the chapter for forming a new team:

NOTE   *If you're refielding a team with the same members as the previous season, skip ahead to "Meet with the Returning Team for the First Time" on page 51.*

1. Announce that the team is accepting new members.
2. Begin accepting applications with a well-defined deadline.
3. Select new team members.
4. Hold the first team meeting.

Announce an Invitation for New Members

As with a new team, the methods for making the announcement are the same. But if the previous season's team members will fill the slots first, the announcement might also include the number of open slots on the team.

You must decide whether returning team members will assume the previous season's roles and duties (see Chapter 6), but you can address that question at the first team meeting. For now, just remember to mention the start date, application deadline, and the application method (interview, essay, and so on).

Accept Team Member Applications

When the application period for new members starts, use the same methods for an existing team with the same caveats as those for applying to a new team.

If returning team members will keep current role assignments (Captain, Building Team Leader, Programming Team Leader, and so on), consider modifying the application process to allow for applicants to apply for specific open positions. This allows applicants to decide whether they wish to apply for the team with the knowledge of which positions are taken and which are open.

There are pros and cons to allowing returning team members to assume previous seasons' roles, and it is much easier on the coach and other team members if you make this decision before the application period begins. On the plus side, team members who are experienced in specific roles require less coaching and management. On the minus side, filling roles with the same individuals doesn't allow others to learn and develop different team skills. If you're not sure what to do, consider postponing the decision, and take a vote at the first team meeting.

As with a new team, create an application that collects contact information and more information about each applicant, such as an essay or checklist of skills.

Select New Team Members

You can use many factors when selecting team members from a pool of applicants. Keep in mind that you don't need to use them all; some factors might not be relevant to your team, and enlisting an assistant coach or using a team mentor can make the process easier or at least faster.

Existing team members can also assist with finding the right mix of skills and interests from the pool of applicants. Of course, the coach must make certain that the current team members make selections based on what's best for the team. If you have a concern that a member may give a friend special treatment, consider removing the names from the applications and letting the team members evaluate each applicant based solely on the application.

Some examples of factors that you can use during the selection phase include the following:

Experience and/or Seniority Due to age restrictions, this may be the last year for some applicants to participate in FLL.

Unique Skills Look at the skills of the returning team members, and try to select new members that fill in any gaps in the team.

Gender Balance If last season's team was an all-girl team, consider giving some preference to boys (and vice versa).

Displayed Interest Again, watch for applicants who have positive energy and interest for the FLL experience. These types of applicants frequently bring new ideas and a curiosity that can take a team to a new level of competition.

Consider a Second Team

Just as when you form a new team, you may find that the number of applicants for an existing team is quite large. If this is the case, the coach may want to seriously consider investigating the formation of a second team. If the interest in FLL is high enough that existing team is overwhelmed with applicants, the coach might suggest this to school officials or parents. However, all the issues that go with supporting an existing team or starting a new team will arise: finding coaches, mentors, and meeting facilities; establishing funding;

obtaining community/parent support; and so on. But if a large enough group of interested students exists, a coach or mentor can enlist their aid to approach their parents, teachers, and community to request the creation of a new team.

Meet with the Returning Team for the First Time

Now that your existing team has filled its roster, it's time to hold the first meeting. You want to try to accomplish the same tasks during this gathering as with a completely new team, and some of these are covered in Chapter 6. Refer to "Holding Your First Team Meeting" on page 47 for information.

Building Your Team

Whether you are putting together a new team with rookie members or bringing new members into an existing team, some of the most important activities your team will participate in occur in the preseason stage. These include getting to know one another, discussing strategies for the competition, and planning the time the team will need to work.

Because preparing a team for competition can involve many aspects, you may want to have two or three team meetings (rather than one initial gathering) before the season starts. In the preseason, many teams schedule a few fun events, such as pizza parties or movie nights, that allow not only the team members but also the coaches, mentors, and parents to talk and plan.

Members can take advantage of the time before receiving the FLL competition kit (mat and model pieces) to work on team-building activities. For example, once you assign roles, test the assigned duties by having the team attempt a mock competition or maybe a previous season's activities. Observe how the members work together, and look for areas where the team may get stuck. Discussing these issues with your team and brainstorming possible solutions is a great way for the team to build its communication and diplomacy skills.

Preseason meetings also offer great opportunities for coaches, parents, and mentors to instruct the team about things such as researching efficiently, building and programming robots, and so on. These skills help prepare the students for the upcoming competition.

Even though you can't research the specific Project topic during preseason meetings, you may use this time to become familiar with the challenge topic so you can spend more time later researching the specific project. For example, in the Nano Quest season, teams could have researched general nanotechnology before the specific Project details were released.

A Team Is Born

Congratulations—you have created an official team! Even though you may not have your FLL competition kits yet, there's plenty to do that doesn't involve the competition or the Retail/Education NXT robot kits. Teams can begin work on the Research/Project portion, brainstorming a community activity (see Chapter 15 for more details), researching good robot designs, and so on, even though they won't be able to work on the actual missions yet.

In Chapter 6, we offer suggestions on organizing your team for the projects ahead, and in Chapter 7, we discuss what it means to work well as a team. Although Chapter 8 is written for coaches, even if you're not a coach, you may find it useful for helping students, parents, mentors, and supporters stay focused, organized, and excited about the upcoming season.

6

MANAGING YOUR TEAM

When it comes to participating in competitions, each team discovers and uses a variety of management methods that enable the team to work and play well together. Options run from total democracy to benevolent dictatorship, but the most successful teams typically find that some sort of middle ground is best for productivity.

In this chapter, we review some examples of team structures as well as the various roles and duties the members might assume. We also discuss making decisions and voting methods, because team members will have differences of opinion. It's good to have some ideas about how to get past a roadblock and begin moving forward again.

Finally, we discuss some methods for tracking a team's progress. Knowing how much work (or how little) a team has completed can help determine whether the team needs to hold more team meetings and help judge whether a team is ready for competition.

Team Structures

Most FLL teams use a combination of one or more of the following structures:

- Team Captain
- CEO and Board of Directors
- We're All Equals Here
- Building Team and Programming Team
- Quick Response Teams
- Chassis Team and Attachment Team(s)
- Robot Team and Research Team

We cover each structure in the following sections, but remember that there is no single best method for structuring a team. Some teams try different methods from season to season until they find one that works. Others discover that one method works well given unlimited time but that they need to use another for making quick decisions.

Team Captain

Sometimes, a team may want to give final decision-making power to a single team member, called the *Team Captain*. For example, consider a team with one older, experienced member and the rest younger, first-year rookies. Given this situation, the rookies might want to give the veteran final say in any decisions and the power to resolve disagreements. All members would be encouraged to discuss options and voice their opinions, but the Captain would make the final decision.

CEO and Board of Directors

Businesses hire Chief Executive Officers (CEO) to manage the organization and hire key personnel. The CEO has a Board of Directors with the power to veto the CEO's decisions and, if things get really bad, remove the CEO. You might consider a similar structure for your team.

A team with a CEO structure chooses to give one team member the power to assign duties, match persons with the proper skills to the right jobs, and make decisions about the direction of the team's efforts. Another group of team members makes up the Board of Directors (if your team uses this method, choose an odd number of members for the board to prevent a tie vote).

WARNING   *While a team will typically survive the overturning of a CEO decision, removal and replacement of a CEO could completely devastate a team. Choose this structure only after careful examination and discussion of the alternatives.*

We're All Equals Here

Just as it sounds, this type of team structure is all about equality. Every team member has an equal voice and vote on all team decisions. While this can be a great way to manage both new and experienced teams, there are some hazards.

For one thing, a full FLL team can have up to 10 team members (not counting the coach). Obviously, it can be difficult getting 10 individuals to agree on anything, let alone every decision during the course of a season. And when decisions can't be made, the team stops moving forward.

Also, when all members have equal say, consider how the team will handle role assignments. For example, what if everyone wants to do the same thing, such as build the robot?

A team choosing complete democracy sounds good on paper, but the reality isn't so easy to implement. Sometimes difficult decisions must be made, and sometimes duties must be assigned to make progress.

WARNING *This type of structure can also prove difficult when team members have a wide range of skills. Many members will have experience with handling certain portions of the competition (programming a robot, for example), and giving inexperienced team members an equal vote in all matters has the potential to ignore the hard-earned lessons (such as robot design or suitable research topics) that veterans have learned in previous seasons.*

Robot Team and Research Team

You might choose to divide the team into a Robot Team and a Research Team. The Robot Team would work mainly on the Robot Game, while the Research Team would work mainly on the Project. This structure allows the members to focus on the part of the competition they like best.

For example, teams often have members who don't feel confident in their building or programming skills. They love doing research and presentations, however, so these team members could be part of the Research Team.

If you decide to split up your team this way, make sure the lines between teams aren't etched in stone so that the team does well in the judging. For example, the Robot Team should participate in the Project presentation, since the judges will evaluate the team based on participation. Judges will also ask the team questions about its Project, and these questions can be directed to any member. Make sure that the Robot Team is familiar with the background and research of the Project and that the Research Team is familiar with the robot's construction.

Building Team and Programming Team

The Robot Game portion of the competition has two major parts: building and programming. Many teams find it helpful to divide their Robot Team into two groups: one responsible for building the robot (Building Team) and the other for programming the robot (Programming Team). Just as with dividing the team into Robot and Research Teams, this division enables members to focus on one aspect of the robot.

Again, this structure is not without its pitfalls. Teams that choose this structure frequently find that the programming team is not able to begin work until the design of the robot is finalized, which gives them less time to refine the program. For example, if a robot is designed to use the Light Sensor, the Programming Team will need to know so it can properly integrate sensor feedback as part of the program. Something as simple as changing the position of the robot's motors can cause major programming headaches if the program expects the motors to be at specific locations on the robot.

If you adopt this structure, be sure that the Building Team and the Programming Team communicate well. The Building Team may have a great looking robot on the table, but if it hasn't shared its design, the Programming Team won't have time to complete a test program for the robot to execute. Likewise, if the Programming Team intends to experiment with a program that uses a third motor to open and close a claw, it needs to let the Building Team know so that it can experiment with claw designs (see "Chassis Team and Attachment Team(s)" on page 57 for a way to divide the Building Team even further).

WARNING   *With many teams, the Programming Team is often smaller in number than the Building Team. The Building Team frequently develops multiple test robots while the Programming Team is still working on its first program. If you are using this type of team structure, try to balance the two teams. If the Programming Team, for example, has one member and the Building Team has three, how might the team cope if the programmer is sick on competition day? Balance allows for not only a sharing of knowledge amongst a larger group of team members, but it can also help teams stay on an even pace so one group doesn't get too far ahead of the other.*

Quick Response Teams

Sometimes having too many team members working on the same robot or program can be detrimental to progress. When multiple robotics kits are available, many teams find it helpful to use small groups, called *Quick Response Teams*, of one to three members. These Quick Response Teams can work in parallel to more quickly tackle building and programming tasks.

Each Quick Response Team is responsible for building and programming a robot. Once each small team completes a design, the entire team can observe, discuss, and make decisions about the different designs. The Quick Response Teams' goal is to develop as many options as possible for the final design.

You can also use Quick Response Teams during the Project portion, with each subteam brainstorming and researching ideas for the Project. The end result should be a larger pool of information.

WARNING   *Quick Response Teams work well for teams with 9 or 10 members and 2 or 3 robotics kits. If only 1 robotics kit is available and the team consists of fewer than 6 members, this structure might not be the best choice.*

Chassis Team and Attachment Team(s)

If you have a large number of Building Team members, consider dividing the Building Team into a Chassis Team and one or more Attachment Teams. One team will design the base, or chassis, of the robot and its mode of movement. The remaining Building Team members design the various attachments the robot will use to accomplish its missions (for more on attachments, see Chapters 10 and 11).

WARNING *As with the Building and Programming Teams, good communication between Chassis and Attachment Teams is required if the robot is to have any chance of successfully completing missions. Be sure that the Chassis and Attachment Teams share information frequently and brainstorm together on the placement and function of all parts of the robot.*

Team Structure Summary

Regardless of how you organize your team, observe how the team works together during the season. Take notes on both the team's successes and failures, information that proves especially useful if the same team chooses to compete in a following season or if the coach organizes a new team.

If your team fails to coalesce and you think the reason is due to its structure, take a step back and try to understand the reason for its lack of progress. Perhaps a change in team structure will help solve the team's particular problem.

FLL Team Roles

During the course of an FLL season, there are numerous jobs to be performed. Some of the jobs are extremely important, and others are not so vital. A key to a happy team, however, is to make certain that every member has one or more duties that have visible and positive effects on the team's successes.

The coach can assign jobs, or the team members can nominate individual members and vote on their selections (secret ballots are helpful to avoid offending anyone).

When voting, allow the team members a moment to speak about why their skills and experience make them suitable candidates for particular jobs. Then have everyone vote.

NOTE *The first team meeting may be the best time to hold this type of vote. If team members prefer to postpone the vote until they get to know one another, the coach could temporarily assign any or all of the roles until a team vote.*

The list below contains several examples of jobs (not necessarily in order of importance) to fill on a team, but your team may find a need for other positions. Keep in mind that for small or large teams, you don't need to go overboard on assigning job titles. Often one role brings enough work to satisfy a team member for an entire season. Assign roles that will benefit the team and satisfy an actual team requirement; if the team doesn't have access to a video recorder, for example, there's really no need to assign someone the role of Videographer.

- Team Captain
- Co-captain
- Building Team Leader
- Programming Team Leader
- Website Manager
- Videographer
- Photographer
- Blogger
- Project Leader
- Program and Data Manager
- Equipment Manager

Team Captain

The Team Captain speaks for the team and helps keep the group focused on the competition. A good choice for this role would be a member with past FLL competition experience.

The Team Captain is frequently called on to make decisions for the team, including settling disagreements. Team members may also agree to allow the Captain to assign other team roles. A Team Captain often works the closest with the coach to arrange meeting locations and times, communicates with other team leaders (such as the Building Team Leader or Programming Team Leader), and acts as spokesperson for the team in the community and other events.

Co-captains

If two members tie in a vote for Team Captain, consider having two team leaders, or Co-captains. Teams may also elect to use Co-captains to eliminate providing any one individual with complete team authority. Co-captains act like the Team Captain except that they must make decisions together.

Some teams choose one male and one female Co-captain; others elect a President and a Vice President with the expectation that the Vice President will become President next season.

Building Team Leader

The robot-building group often designates a Building Team Leader who is responsible for managing the work on the robot as well as communicating the team's progress to the coach and Team Captain. The Building Team Leader should be skilled in the building of MINDSTORMS robots and be able to review robot designs for durability, ease of use, and proper functionality when it comes to the Robot Game missions. She should also take the time to train less-experienced members and to educate the team on the final robot design, making certain that each member can speak about the team robot.

Programming Team Leader

The Programming Team Leader manages the programming team and ensures that programs are being developed that work well with the robot design(s) from the Building Team. He should be familiar with all aspects of the NXT-G programming language, be able to quickly debug programs, and have good communication skills. The Programming Team Leader is also responsible for ensuring that all team members learn about the program(s) used in competition and that each can speak about the robot's program(s) and functions.

Website Manager

Many teams enjoy having a website as a method to share details about the team, its members, and community projects. The website is also a great place for parents and sponsors to monitor the team's progress.

The Website Manager is responsible for updating the website with pictures, video, and write-ups about the team's work. During competitions (especially if the team travels), she can help keep the community informed about game results as well.

The Website Manager should ideally be skilled in website design and maintenance.

Videographer

Some teams choose to record video of their gatherings and competitions. Assigning one or more individuals the responsibility of documenting the season using a video recorder can allow a team to put together an end-of-season DVD to share with friends, family, and sponsors. Video recorders are often expensive, so be sure that the Videographer is diligent about storing and protecting the equipment when traveling.

Photographer

Like the Videographer, a team often has several people take photos during team meetings and competitions. A team's Photographer can coordinate with parents and other teams to swap photos. At the end of the season, he can create an online photo album (together with the Website Manager) or work with the Videographer to create a DVD with videos and photos. The Photographer may also be responsible for providing pictures about team events to local newspapers.

Blogger

Like the Website Manager, a team's Blogger writes about the team's activities and posts the articles on a web log (blog) or the team's website. Blog updates should be short and sweet, with an emphasis on current events. Readers should be able to read a quick paragraph or two about the team and its activities.

After a team meeting, for example, the Blogger might go home and post a short write-up about what the team did that day. Blogs are a great way to document progress and share that information with other teams, family, and friends.

Project Leader

The Project Leader helps keep the Research Team organized. For example, she might work with the coach to schedule field trips, interviews, presentations, and any other activities associated with the Project. She could also help assign research to members of the Project team and make sure that the work is progressing steadily.

Program and Data Manager

During the competition season, the team generates a lot of information, both from its Project research as well as from the robot building and programming work. The Building Team should frequently photograph its robot designs (or use CAD software to create graphical building steps, as discussed in Chapter 11). Keep these both for historical value and safety in case the team's robot is damaged and the team needs to build a new one. Likewise, the Programming Team will generate a lot of test programs as well as the final programs used by the robot, and they should be carefully tracked and saved.

The Program and Data Manager is responsible for copying all team data to a central location or making backups of the data (on DVD or CD, for example). This duty is crucial to ensuring that the team doesn't lose any of its hard work.

You could further break this role down into a Building Backup Manager (for the robot) and a Data Recovery Manager (for all programs and Project files), with the corresponding members each taking care of one category.

Equipment Manager

Whether scheduling a team meeting or traveling to a competition, there are laptops, PCs, flash drives, cameras, video recorders, and many other items besides the robot that should be on the team's inventory list. The Equipment Manager's job is not to own or keep all this equipment, but to coordinate with the coach, mentor(s), and team members to ensure that resources and supplies are available.

In the rush to get to meetings and competitions, many teams forget crucial items (such as batteries or even a laptop), but the Equipment Manager can save the day by making sure that calls are made and emails are sent to remind people to bring the necessary equipment and supplies.

Having Fun

FLL should be a fun experience for all involved. Don't let the team roles and duties overshadow the fact that the team should be learning and playing together as its work progresses. Team disagreements during an FLL season will always occur, but hopefully they will concern items such as the robot design, program structure, and research topics. If you find that team conflicts revolve around the assigned roles, you may want to consider the following to resolve these disputes:

- If team members find that their duties are stressful or tiresome, consider assigning two or more members to each role or job. Sometimes simply having someone to help can make all the difference.

- Don't let a role go to a team member's head. Make certain that the team members are all recognized for their jobs and the contributions they bring to the overall team success.

- Assign second-in-command roles to all key team positions. If the Building Team Leader is out sick for a meeting, for example, the Building Team Lieutenant can lead the Building Team during that meeting. If disagreements cannot be resolved between team members or by the Team Captain, inform the coach. Be sure to give the coach privacy and time to address any situation. Team members should never make hasty decisions without first consulting the coach.

- Have an occasional "no roles" meeting that the coach moderates. During this meeting, encourage the team members to discuss the current situation as well as any other concerns related to the robot or Project (however, this is not a meeting to point out problems with teammates or to complain about another's work).

- Cross train members when time is available; if the Programming Team has some free time, for example, consider having those members assist the Project Team by either acting as judges (to let the Project Team present their material) or accepting assignments to research specific topics. The Building and Programming Teams often find it beneficial to observe and ask questions of the other team.

Making Decisions and Voting

No matter what team structure you choose or what roles are assigned, every team faces decisions during a season that you need to resolve *as a team* to move forward. Sometimes decisions can be extremely simple: "Pizza or hamburgers for Saturday's gathering?" Others might not be so easy: "The Building Team has two robot options, one with wheels and one with tracks. Which should we use?"

Number of Members Needed for a Vote

One decision a team can make early in the season is whether you can only hold votes that affect the entire team when all the members are present. If the team agrees to let those in attendance (when a vote or judgment is needed) make a decision, team progress will be much smoother and faster.

Requiring all members to vote on a decision will certainly slow a team down; imagine what the result will be if three or more votes are needed at each team gathering. The team will be lucky to get any work done if discussion occurs before a vote, and the vote and tally can be time-consuming as well.

Final Say

Another item that team members can agree on before the season starts is the issue of final say. Sometimes voting will result in a tie; other times there aren't enough members present to make an informed decision. When this situation occurs, it can be beneficial if someone has been given the power to make a quick decision (sometimes called a *command decision*). You may want to give a Team Captain or Co-captains authority to make a command decision when one is required.

Voting

There are various methods for taking a vote, but don't let your team get bogged down in the details. Two of the most popular methods are also the simplest: *secret ballot* and a *show-of-hands* vote.

Secret ballots can be as simple as the team members writing their votes on index or note cards and the coach counting the results. When multiple decisions are required, consider writing each item on a whiteboard or poster and numbering it. When the team members write down their votes, they number the decisions so there is no confusion about which vote corresponds to which question (see Figures 6-1 and 6-2 for a sample).

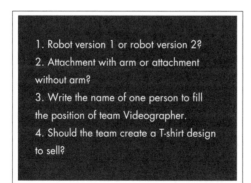

1. Robot version 1 or robot version 2?
2. Attachment with arm or attachment without arm?
3. Write the name of one person to fill the position of team Videographer.
4. Should the team create a T-shirt design to sell?

1. I vote for robot version 2
2. Attachment without arm
3. Marie LaBolle for Videographer
4. Yes

Figure 6-1: Write down and number multiple decisions for voting.

Figure 6-2: Team members should number their voting decisions on a card.

The show-of-hands voting method is best used for yes or no decisions. The coach or Team Captain asks the question (or writes multiple questions on a blackboard) and asks the members to raise their hands for *yes*. Those who do not raise their hands are voting *no*.

The Team Captain should break ties if he or she received the ability to make command decisions.

Here are some final suggestions and tips for making decisions:

- After you take a vote, remind the team members that it's time to get back to work. Don't let them get bogged down by a vote that doesn't go their way.

- If a team is split on an issue and cannot make a decision, examine how much time it might take to consider both views (two robot designs, for example). It's always best to make a decision and move forward, but sometimes it might be in the best interest of the team to hold off on a vote and allow a little more time to investigate all options.

- A team may find it best to allow subgroups (such as the Programming Team) to vote privately on certain decisions that affect them more directly, rather than putting the vote to the entire team. The Building Team, for example, might be given authority to make any decisions that affect the robot design.

- Before taking a vote, give team members of opposing views time to explain why they think their idea is best. In fact, simply discussing an issue may result in a unanimous agreement, and a vote won't be necessary.

Team Progress Tracking

We close this chapter with a short discussion on tracking a team's progress during the season. Teams often feel overwhelmed at how much work they need to do at the start of a season, from registering the team, to finding sponsors, arranging meeting places and times, testing the robot, and so on.

The list of activities is a long one, but like the old saying goes, "How do you eat an elephant? One bite at a time."

NOTE *In Chapter 8, we discuss software that can help a coach and team manage their time, arrange meetings, and track work completed.*

Start simple. The coach or Team Captain may want to start with a list of all the activities to complete. Write down any deadlines, and jot down the name(s) or team(s) that will be associated with each activity. The goal will be to transform your To-Do list from paper into something that is visible and motivating to the team.

The Gauge as a Progress Measurement Tool

When an organization is running a fund-raiser, they sometimes use a *fund gauge* similar to the one shown in Figure 6-3.

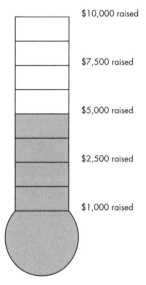

$10,000 raised

$7,500 raised

$5,000 raised

$2,500 raised

$1,000 raised

Figure 6-3: A fund gauge shows
how much money the team has raised.

Manage your team's primary tasks similarly by creating a *progress gauge*. The concept is simple: Provide a visual measure of the team's progress during the season, as shown in Figure 6-4.

Create your team's progress gauge on poster board, and label the key activities. As you progress up the gauge, you'll find that your team receives immediate satisfaction and motivation at its first gathering when they notice that the first few items have already been completed: register the team and hold first team meeting. To show a bit more progress on the gauge, add items such as "assign team roles" or "break into subteams." The key is to update the gauge constantly to show finished work.

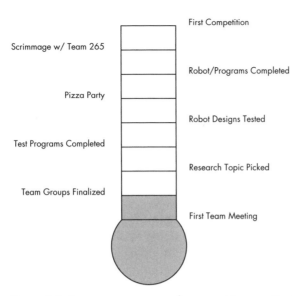

Scrimmage w/ Team 265

Pizza Party

Test Programs Completed

Team Groups Finalized

First Competition

Robot/Programs Completed

Robot Designs Tested

Research Topic Picked

First Team Meeting

Figure 6-4: A progress gauge can be a great team motivator.

You might also consider creating separate progress gauges for the Robot Team(s) and Research Team. All the team members can then determine at a glance how far along each subteam is in its work.

One final suggestion for your progress gauges: Be sure to add some noncompetition activities along the way as rewards for completing certain tasks. For example, when the Building Team finalizes the robot design, consider adding "Robot Completion Pizza Party" to your progress gauge. When the party is over, color in one more item on your gauge, and the team will be highly motivated to reach the next noncompetition activity.

Chapter 7 provides you with some ideas for working well together as a team. The FLL season is long and it can seem like forever before a competition rolls around. Some real-world advice is exactly what you need right now, so we share some tips and stories from teams that have been there and made it through. Know that other teams have been through the same trials and made the same mistakes, and you're not alone.

7

THE TEAM EXPERIENCE

This chapter examines ways that a team can enrich not only its own FLL experience, but also the experience of all the teams it interacts with during a season. We also share feedback from real teams (scattered throughout the chapter) about how they organized, worked, played, and competed.

This chapter isn't as much about what team members will receive from the FLL experience as it is about what the participants should strive to give back. Every team member has a responsibility to the team; they support and help one another in an effort to make FLL the best experience it can be for all involved.

Encourage Your Team

FLL teams do not last forever; in fact, chances of the exact same team members being around next season are slim. Members move away or get too old to compete; new members join, and sometimes the team has a different coach.

Take the time to appreciate and encourage your team members and the diverse skills and personalities that make your team different from every other FLL team.

Team members should encourage one another in whichever endeavors they choose to tackle. If your Programming Team does a great job of debugging the program for Mission #3, tell them so. If the Research Team picks a topic that is interesting and inspiring for the community presentation, someone can see if the local newspaper will cover the event.

If a teammate goes above and beyond what is expected, let her know that you appreciate it and that it will benefit the team. Often so much is happening at team gatherings, the coach and team members can miss a lot. Team members should always look for ways to remind the group about how much fun they're having.

TEAM PERSONALITIES

A team often has its own "personality" that its members can recognize when it is being described. FLL team Central SpaceLab One from Indiana shares this story:

In 2006 we won the Director's Award at the World Festival in Atlanta. We figured out we had won before our name was announced from the description given by the announcer. Some of our team members were jumping up and down with excitement, waiting to hear the team name called. It was a great moment that capped several months of hard work.

Encourage and Share with Other Teams

One of the most exciting activities that two teams can share (besides the Robot Game) is simply getting to know one another. If you think your team is fun, wait until two teams get together . . . or even three or four!

The FLL team members' shared interests in robotics, math, and science breaks down barriers. There really is no better icebreaker than two teams sharing their experiences and displaying the results of their hard work.

Because FLL is international, teams can often communicate with teams from around the world, sharing photos and information about their countries. And when teams that get to know one another qualify for the World Festival, meeting face-to-face is one of the most rewarding opportunities a team can have.

Share Your Experience

Share the FLL experience with your community, not just your parents and friends. Your community is full of organizations and individuals who may also enjoy the experience, so let them know! Here are some ideas for doing so:

- Contact the local newspapers and television stations to inform them of your team's work and any upcoming competitions.

- Post flyers in coffee shops, grocery stores, and the post office inviting the public to attend the next competition and cheer on local teams.
- Invite retirement communities to come to watch your team compete.
- Ask the fire department if it will bring a fire truck to the next competition, and have some firemen talk about its functions and their job.
- Send letters to your parents' employers inviting them to the competition.
- Ask elementary schools if your team can present information about FLL to get the younger kids excited.

INTERACTING WITH OTHER TEAMS

I've attended FLL World Festival since 2006. I've been fortunate to meet teams from Singapore, Germany, Turkey, Israel, Japan, China, Canada, Taiwan, Mexico, the Netherlands, the United Kingdom, Spain, France, and more. Speaking to these teams (some through interpreters) about their robots, their projects, and their countries was an opportunity I couldn't resist. I would suggest that every FLL team locate a team from another country and communicate during the season; try to learn a little more about its culture and its studies as well as find some common areas of interest.

—JFK

Give Back

As a team, every member has the opportunity to give back to schools, community, religious institutions, and other organizations that have provided support. Giving back may be as simple as having every team member send thank-you cards with notes of gratitude, or teams might consider volunteering to do some lawn work or a car wash to help raise funds for sponsoring organizations.

OPPORTUNITIES ABOUND FOR FLL TEAMS

I can't say enough about the positive impact FLL has had on our team members and their families. I've seen such personal growth in my team members this year—the confidence, the knowledge that today's youth can make a difference, the deep friendships, and the increased understanding of how different people function in different types of situations. Perhaps more than anything, my team members have learned that they can open up unique opportunities for themselves. During a visit to a museum, someone walking by asked the students about their T-shirts, and a lengthy conversation developed from there, resulting in a unique workshop opportunity for the kids. These kinds of things seem to happen more and more often for my team. We have had so many doors open up to us because the kids are so passionate about their team and learning.

—Christine Bibic, Head Coach for the W.A.F.F.L.E.S. #87 Core Values Team from the World Festival

Carefully consider the financial support, time, and equipment that your team receives. If the sponsoring organizations do not wish to receive compensation (whether volunteer work or something else), consider having your team choose an unaffiliated nonprofit organization to assist as needed. For example, instead of compensating a sponsor, your team might organize a donation drive for a local after-school program that needs supplies or funding.

An FLL team is a powerful thing: a group with high motivation, intelligence, and creativity, which can do amazing things when its members put their minds to it. Multiply that team by more than 10,000 teams with similar intentions, all helping out and giving back in their local communities, and you have an international movement of students that is powerful on both the competition field and in the world.

GIVE BACK AND HELP FLL GROW

Our team hosts a summer camp for children 6 to 11 to teach them about robot design and programming. We do this as a fund-raiser to support our team expenses and as a service to our community. It helps the team members practice leadership and mentorship while exposing young children to robotics and the concept of LEGO League. We are sure to provide information and resources to the campers and families about how to start their own team.

—Team μ (the Greek letter *mu*), Tampa, Florida

Methods for Sharing Your Experiences

While writing this book, we've been lucky to see and hear how teams share their FLL experience with friends, family, and other teams from around the world. We collected some of those experiences and list them here, but you'll find still more in the forums at *http://www.thenxtstep.com/smf/*. Click **Book Discussions** and then **FLL: Unofficial Guide** to read the experiences of others and share your own.

Websites and Blogs

Consider tracking your team's progress through the season using a website or blog. Share photos and video of your team in action. To liven things up a bit, you might even consider sharing some of your failed programs or building designs that don't make the cut. Make your site fun and entertaining, and give friends and families the URL so they can read about your team meetings, competitions, and research.

ALL-GIRLS TEAMS BREAK STEREOTYPES

Team members from three teams fielded by the Atlanta Girls' School (AGS) shared their thoughts with us on FLL in general and competing as an all-girls team in particular. The veteran team members (Megan, Izzy, Victoria, Patrice, and Emily S.) adopted an informal motto of "Have fun and don't take things so seriously." They stressed communication and focused on keeping stress levels down. The rookie team members (Leslie, Emily R., Kierston, Africa, Sophie, Devin, Jordan, and Ferra) used team-building exercises and enjoyed being on an all-girls team because they felt it was easier to fit in. Both veterans and rookies agreed that keeping the entire team informed of all activities was important. Some of their suggestions include the following:

Circle Time   To resolve conflicts, the all-girls teams used this brainstorming technique in which no one can leave until a disagreement is resolved or a decision is made.

Democracy Always   Both veteran and rookie groups agreed that teams work and communicate best when everyone gets a vote and all opinions are respected.

Flaunt Your Differences   The AGS teams enjoyed the uniqueness of sending all-girls teams to many competitions. The young ladies used this to their advantage and showed off their teams' personalities, but in a fun, nice way.

—Teams: Women in Black, Questionmarks, Nutty Acorns, Atlanta Girls' School (AGS), Atlanta, Georgia

Diaries and Journals

Document your team's progress in a diary or journal throughout the season. Encourage all team members to record their experiences, including road-blocks, lessons learned, and breakthroughs. Supplement the journal with pictures of the team's robot, screenshots of the programs, and any other materials that liven up the writing. Consider donating the finished work to the school library for future teams to read and learn from. Be sure to include a team picture as well as advice and encouragement for future teams.

Photo Albums

Create a photo album to give to the coach or multiple albums for team mentors and sponsors. When taking photographs for the album, adhere to one rule: Take pictures of everything, including all team gatherings, competitions, and any place that the team assembles.

If possible, include simple captions for your photos, listing the names of the pictured people and a short sentence about what they're doing. As with a diary or journal, consider donating the finished album to the school library.

Team Posters and Promotional Items

Your team might consider commemorating the season by providing each member with a keepsake that can bring back memories of the fun and excitement of FLL. Examples of these items might be T-shirts or posters.

At most FLL competitions, you will see students wearing some sort of team shirts. These shirts usually picture graphics or signatures that were created by one or more team members. The team name and number, as well as a team slogan, are also often included.

Some teams upload team graphics to sites such as *http://www.cafepress.com/*, where team members can create custom items such as shirts, hoodies, book bags, and more. Others use funds from sponsors to print team shirts in quantity. (Shirts are a simple way for team members to distinguish themselves and for them to find one another at competitions among dozens or hundreds of other participants.)

Some teams also create an end-of-season poster that includes a team photo. Team members sign one another's posters (like a yearbook) to hang at home as a reminder of the season.

The FLL Journey

We hope that you have some ideas for adding to the FLL experience; although your team will absolutely enjoy the Robot Game, Research portion, and competition environment, always keep your eyes open for ways to exceed your team's expectations.

The FLL season moves fast, so constantly remind the team members to write down their thoughts and observations. Ask them to be diligent in photographing robot designs, team gatherings, and competitions. Ask the members and their families to record video of as many activities as can be captured.

FLL truly is a journey, so make sure to record as much of the trip as possible. And when the season is over, your team will be able to relive the experience as often as it likes.

8

COACHING A TEAM

In this chapter, we offer some advice that will hopefully clear up any misunderstandings about what a coach *should* and *should not* do during an FLL season.

Opinions from previous FLL coaches vary from "Not too bad—the kids did all the real work" to "Herding cats is an easier job." The ideal experience for a coach will ideally fall somewhere in between. Most coaches probably agree, however, that coaching is one of the hardest jobs you'll ever love.

During a competition season, coaches often find that the task of scheduling team meetings and competitions becomes a job in itself. Communicating with students, parents, and sponsors regarding setting up meeting times, coordinating rides, and simple fund-raising activities can consume much of a coach's time, so we offer some suggestions for software and online applications that are sure to ease a coach's stress (a bit) and bring some peace of mind.

Finally, find real-world advice from coaches scattered throughout the chapter. By the time you finish this chapter, we hope you'll breathe a little easier. Let's get started.

The Coach's Responsibility

There are duties that a coach should reasonably be expected to contribute to an FLL team, and there are actions that fall outside a coach's responsibility.

The FIRST LEGO League *Coaches' Handbook* sums up the division of labor between coach and kids with this statement in "Coach's Promise" on page iii: "The kids do . . . all of the programming, research, problem solving, and building. Adults can help them find the answers, but cannot give them the answers or make the decisions."

Remember that FLL is about the kids, not the coaches. Of course, it's great when a coach learns something new, but team energies should always be directed at expanding and enriching each student. Give your team room to dig deeper, go off on tangents, and learn. For the Robot Game portion of the competition, coaches should resist offering help with programming and testing the robot. Coaches, mentors, and parents can give general instruction but not specific solutions. The coach should invite mentors to speak to the team, pointing to resources such as books or websites, but when asked whether a particular program or technique will work, a good coach's answer should be, "I don't know. Let's try it!"

NOTE *Coaches should encourage students to build test robots based on designs from books and the Internet, but the best source of building instruction is in the NXT-G software. Coaches are encouraged to do their own research so they, too, become knowledgeable about the chosen Project and Robot Design and can provide encouragement when the team gets stuck on a problem. Team members will get frustrated when you don't provide answers, but the feeling of satisfaction when they find the answers is always greater when they realize their success is all their own.*

Perhaps most important, team members should make all decisions on their own, without their coach. This includes voting on items such as Robot Design or the Project's subject. Coaches can provide an opinion if asked, but not specific solutions to problems. Coaches should remember, though, that their opinion may carry more weight among the team members than the opinions of the other members, and they should always encourage students to make their own decisions. If asked for a decision on a particular topic, coaches should in most cases reply, "As coach, I cannot make that decision for you."

The *Coaches' Handbook* lists FLL values that all team members should adhere to—and not just during the FLL season. These values can benefit team members not only during competition, but also throughout their lives. These values are the following:

- Respect each other in the best spirit of teamwork.
- Behave with courtesy and compassion for others at all times.
- Honor the spirit of friendly competition.

- Act with integrity.
- Demonstrate gracious professionalism.
- Encourage others to adopt these values.

Encouraging a team to honor these values is the coach's most important duty. Respecting others, behaving appropriately during and outside team gatherings and competitions, making ethical decisions, and similar behaviors are all things the entire world could benefit from.

A coach can best judge a team's success at the end of the season not only by the number of trophies won but also by the impact the team had on its members, other teams, and the community that supported it. You hear the phrase *gracious professionalism* a lot during FLL competitions. Many teams adopt this phrase as their motto, recognizing that FLL is not just about competition; it's a way of being. Coaches should understand the FLL values, integrate them into their management style, and remind the team members always to remember and adhere to the values.

Online Tools for the Busy Coach

Understanding the rules and values that a coach must abide by is important, but how can you manage all of the day-to-day activities that a coach needs to attend to, such as scheduling the next meeting, arranging transportation, finding volunteers to help with the fund-raiser, and so on? A coach's work is never done.

Fortunately, the Internet is full of useful tools that can simplify a coach's life and help keep the team organized. This chapter briefly covers Doodle and Google Calendar and lists some additional resources.

Doodle

Scheduling FLL meetings and events can be a real challenge because you must consider the schedules of the team members, parents, mentors, and an assistant coach or two. Thankfully, Doodle (*http://www.doodle.ch/*) makes it easy.

As of this writing, Doodle is a free online tool for scheduling meetings or appointments. After setting up an account, use Doodle to select possible meeting times for an event. Doodle emails your invitees with proposed dates and polls the members for their choices. Once all of the votes are in, Doodle picks the day and time that works for most people. Here's how it works in a nutshell:

1. Create a poll by giving it a name. Examples are "Best time for Pizza Party?" or "Time to gather for Saturday's scrimmage?"
2. Fill in details such as the purpose of the meeting, and provide your name and email address.

3. Select the month and year for your event. Figure 8-1 shows a sample with possible dates highlighted.

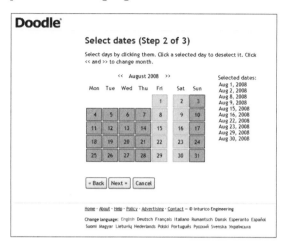

Figure 8-1: Select the days that are available for your event.

4. Enter time options. (For each day you select as an option, you can provide up to four times.)

5. You will receive an email with a link to forward to the invitees. When the invitees click the link, they will arrive at a screen like the one shown in Figure 8-2.

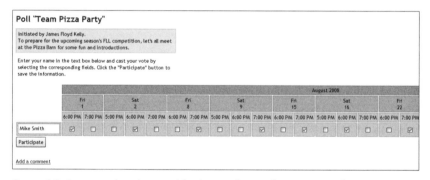

Figure 8-2: Invitees select the possible days and times they can attend.

6. As invitees respond and provide their availability, Doodle offers the best day and time along with the largest number of attendees (Figure 8-3).

As the poll administrator, you will receive an email every time an invitee responds to the poll. Also, as new participants submit their information to the poll, they can view the most popular days and times selected by previous respondents; this is a helpful feature that can encourage visitors to select days and times that are already showing good potential.

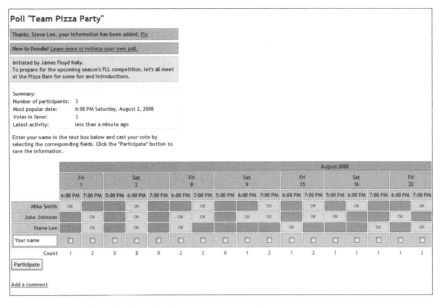

Poll "Team Pizza Party"

Thanks, Steve Lee, your information has been added. Fix

New to Doodle? Learn more or initiate your own poll.

Initiated by James Floyd Kelly.
To prepare for the upcoming season's FLL competition, let's all meet at the Pizza Barn for some fun and introductions.

Summary:
Number of participants: 3
Most popular date: 6:00 PM Saturday, August 2, 2008
Votes in favor: 3
Latest activity: less than a minute ago

Enter your name in the text box below and cast your vote by selecting the corresponding fields. Click the "Participate" button to save the information.

	Fri 1		Sat 2		Fri 8		Sat 9		Fri 15		Sat 16		Fri 22	
	6:00 PM	7:00 PM	5:00 PM	6:00 PM	7:00 PM	6:00 PM	7:00 PM	5:00 PM	6:00 PM	7:00 PM	6:00 PM	7:00 PM	5:00 PM	6:00 PM	7:00 PM	6:00 PM	7:00 PM
Mike Smith	OK			OK			OK			OK		OK		OK			OK
John Johnson		OK		OK		OK	OK		OK	OK		OK	OK			OK	
Steve Lee		OK		OK		OK					OK				OK		OK
Your name	☐	☐	☐	☐	☐	☐	☐	☐	☐	☐	☐	☐	☐	☐	☐	☐	☐
Count	1	2	0	3	0	2	2	0	1	2	1	2	1	1	1	1	2

Participate

Add a comment

Figure 8-3: Doodle calculates the best day and time for the event.

While Doodle cannot guarantee to find a day and time that works for every invitee, it does give a coach the ability to pick the date that will have the most participation without having to make lots of phone calls or send numerous emails.

OTHER SCHEDULING TOOLS

Doodle isn't the only scheduling software game in town. Two other free online tools that offer similar features (with a few differences) include Presdo (*http://www.presdo.com/*) and When2Meet (*http://www.when2meet.com/*). Both services allow you and invitees to select from available dates and times. Try them all and see which one you like best.

Google Calendar

Another excellent tool for coaches is Google Calendar. Coaches can use it to create an event schedule and make it publicly available or restrict access only to invitees. Google Calendar has a ton of features, and although we can't cover them all, we do want to cover a few basics that will get you started.

Adding Events

1. Sign in to Google Calendar at *http://calendar.google.com/* (if you don't already have a Google account, click **Create an account** and follow the instructions).

2. When Google Calendar opens, you will be viewing it in Week mode. Also available are Day, Month, 4 Days, and Agenda (click **Help** in the upper-right corner of the screen for more information).

3. Press the M key on your keyboard; this changes the calendar to the monthly view (Figure 8-4). For our examples, it will be easier to work in this view.

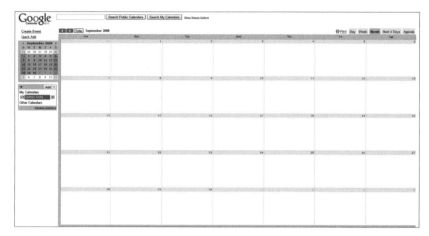

Figure 8-4: Monthly view in Google Calendar

4. To enter a new event, click a specific day. This opens a small box in which you can type a description of your event (Figure 8-5). Click the **Create Event** button when you're finished. (If you press Q instead, a smaller box will open, allowing you to enter event information; type the date and time and it will place the new event accordingly.)

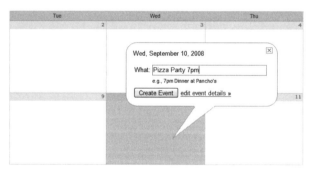

Figure 8-5: Enter your Google Calendar event.

5. The new event will be entered on the calendar. In your description, you can specify the time and it automatically enters the event at the appropriate time in Day view.

6. Navigate forward and backward in any view (Day, Month, and so on) using the N and P keys (*next* and *previous*), and enter additional events as needed.

Sharing Your Calendar

Once you've created the calendar, you can share it with team members, parents, mentors, and sponsors. Anyone wanting to view the calendar also needs a Google account; you have an opportunity to invite them (using an emailed link) to sign up, view your calendar, and even add items to the calendar. The following tells you how:

1. On your Google Calendar web page, click **Settings** in the upper-right corner of the screen.

2. The Calendar Settings window will appear, showing the General screen. Click the **Calendars** link.

3. Click **Share this calendar**.

4. On this next screen (Figure 8-6), you need to enter the email address of each person who will have access to this calendar. For now, leave the drop-down menu option as See all event details, and click the **Add Person** button (this is where you can give someone access to make changes to the calendar by selecting **Make changes to events**).

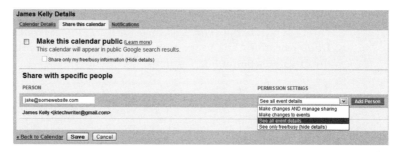

Figure 8-6: Enter an email address to let an individual view your calendar.

5. Continue to add email addresses (clicking **Add Person** after each) to add more individuals. Click the **Save** button when you are done.

6. An alert message will then inform you whether the persons added should be invited to create Google Calendar accounts. When you click the **Invite** button, they will be sent invitations to sign up and view your new calendar.

And that's it! You now have a calendar that parents, team members, and anyone else you care to invite can view. It's a great tool for making sure that everyone knows about scheduled team meetings, competition dates and times, and more.

Once you create a calendar, there are some really fun and useful things you can do with it. Two of our favorites are as follows:

- For a new event, enter an address that includes a zip code. When you or a visitor to the calendar click the event and the small **Map** link, Google Maps will allow you to print a map to that location!

- From your cell phone, send a text message to GVENT (48368) with a short description and date (a time is optional), and your calendar will add the event and send you a reply text message that the calendar has been updated. This is a great feature for adding items on the fly when you're not at a computer. (Keep in mind that this only works if your cell phone plan supports text messaging to short codes, and standard text message rates will be added.)

OpenProj

If you're not familiar with software such as Microsoft Project, you might be surprised at the features these kinds of applications offer. Project management software allows you to enter team project tasks (such as First Team Meeting or Test First Robot Design), set a start/end date or task duration (five days, for example), and track milestones and key deliverables for your team. You can also build in set dates (such as Regional Competition, November 8) and define tasks that the team must complete before other tasks (called *dependent tasks*).

By configuring your FLL season schedule using project management software, you can determine at a glance if your team is on schedule. Our choice is OpenProj (*http://www.openproj.org/*), shown in Figure 8-7. Click the **Demonstrations** link to see some examples of how to use OpenProj and click the **Documentation** link for help.

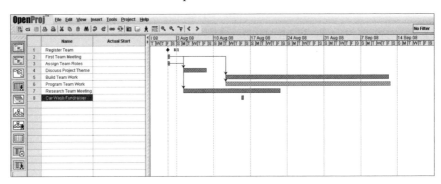

Figure 8-7: OpenProj lets you manage team tasks all season.

Organizers Database (Tracking Donations and Contributors)

Our last application, Organizers Database (ODB), is a simple tool that allows you to track donations and contributor information. After installing the software, simply add contributors to the database using the **New** button (as Figure 8-8 shows). You can include contributor information such as mailing address, email, phone numbers, individual or organization name, and more.

When a business or individual makes a donation, use the **New Payment** button to append the record with information on the amount contributed and the date of their contribution.

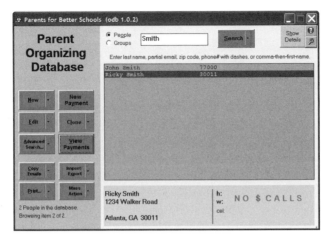

Figure 8-8: Organizers Database allows you to track donations.

Organizers Database can also track donations of time or equipment. At the end of the FLL season, the software makes it easy to filter and print out names and addresses. The team members can then send thank-you cards to the people who've supported them.

You can download Organizers Database at *http://organizersdb.org/home* for free. A user's guide that provides information on additional features also comes with the download.

More Help for the Coach

Finally, make certain to check out the resources listed in the appendix. We include websites, books, and more—all useful to a coach (and team members in general). Be sure to visit some of the forums mentioned; they're a great place to post questions and "meet up" with other teams to share advice.

9

NXT VS. RIS

All FLL robots must be built entirely of LEGO parts, a limitation that makes it easier to build a competitive robot, because all teams use the same starting materials. Using LEGO parts to build the skeletons of robots is very easy compared to other systems (such as Vex) because there's no need to cut pieces to size, weld, and fit parts together.

When building robots to compete in FLL, the LEGO MINDSTORMS robotics system acts as the basis for robots. *MINDSTORMS* is a department of LEGO that makes child-friendly robot equipment. You can learn more about it at *http://www.mindstorms.com/*. The MINDSTORMS system is powerful yet simple: You don't need to know anything about circuits, hardwiring, or advanced programming to build a robot with it. You won't have to weld or solder either; things pretty much snap together, which is the point. This is the first step in the FIRST series of competitions; you'll have the chance to cut and weld later on in the FIRST Robotics Competition.

MINDSTORMS sensors and motors connect to a microcomputer via cables with standard connectors. The microcomputer acts as the robot's brain by receiving information from sensors, performing calculations, and controlling motors. It can be programmed using a graphical user interface and child-friendly language.

LEGO MINDSTORMS now includes two robot systems: the older Robotics Invention System (RIS) and the newer NXT, either of which you can use in the competition.

In this chapter, we discuss and compare each system. If you already know which system you're using, you might choose to skip ahead or quickly run through this chapter to review the components of your system.

Robotics Invention System

The RIS was the first system used in FLL and was the only available option for many years. RIS components consist of electrical parts and construction pieces. The electrical parts include a microcomputer (called the RCX), sensors, and motors.

NOTE *As an interesting trivia fact,* RCX *stands for* Robotic Command Explorer.

There are four types of sensors: touch, light, rotation, and temperature. The construction pieces can consist of any LEGO parts, but the electrical parts are mechanically designed to work most naturally with a mix of TECHNIC *bricks* and pieces such as pegs and gears. TECHNIC pieces are more advanced pieces that help build more complex creations (like robots!). Figure 9-1 shows an example of a TECHNIC brick.

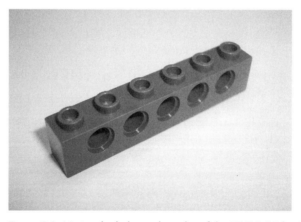

Figure 9-1: Notice the holes in the sides of this TECHNIC brick, where pegs or axles can snap in.

The RIS includes a graphical programming language called RCX Code. To program, a user connects several command "blocks" to form a program. For example, if a user wants to program a robot to move forward and then backward, he might connect a *Forward* block to a *Backward* block. Programs can be downloaded to an RCX via an infrared transmitter. Figure 9-2 shows an example of a program made using RCX Code.

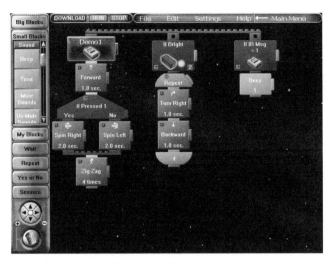

Figure 9-2: An example of an RCX Code program

NXT

The NXT system was released as an update to the RIS in 2006, and FLL has used it since the Nano Quest season. The NXT's counterpart to the RCX is the NXT Brick, which is the microcomputer that acts as the robot's brain.

The NXT Brick works similarly to the RCX, but it has additional features and functionality. One significant added feature is the ability of NXT Bricks to communicate via Bluetooth with each other or with computers. Another addition is the ability to download programs using a USB cable.

The NXT system also includes new electrical parts such as redesigned Light and Touch Sensors, as well as new Sound and Ultrasonic Sensors.

As with the RIS, the NXT system is compatible with all LEGO pieces. However, the NXT's electrical parts are designed to work most naturally with studless building pieces. These don't have those little round "bumps," called *studs*, that you see on LEGO bricks. They include pieces such as TECHNIC beams, gears, pegs, and so on. Figure 9-3 shows an example of a TECHNIC beam.

The official programming language for the NXT system is called NXT-G. Like RCX Code, it's a graphical language that uses drag-and-drop command blocks. However, unlike RCX Code, NXT-G has a different interface and additional capabilities that allow the use of new electrical parts in programs. This brings us to a basic question: Which one should you use—the RIS or the NXT? The answer is not straightforward, so the following sections compare many of the main attributes of RIS and NXT.

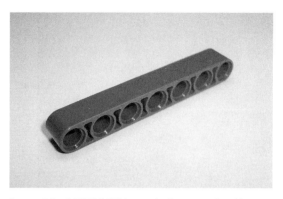

Figure 9-3: A TECHNIC beam looks somewhat like a rounded TECHNIC brick without studs.

The Bricks

Both the RIS and NXT microcomputers have *input ports* for connecting sensors and *output ports* for connecting motors. They also have buttons and display screens that allow users to interact with the microcomputer. Both microcomputers can be programmed to control a robot's behavior by turning motors on and off, receiving and acting on sensor data, performing calculations, and so on.

The RCX

Figure 9-4 shows an RCX, the microcomputer for the RIS. It has three input ports and three output ports. The input ports are numbered *1* through *3*, while the output ports are lettered *A* through *C*. The RCX screen displays basic information such as a program number and data from a sensor.

There are four buttons on the RCX: On/Off, Run, Prgm, and View. The On/Off button turns the RCX on or off; the Prgm button switches between available programs; Run starts the selected program; and the View button causes the screen to display data from a port (each time you press View, the screen displays data from the next port).

A basic speaker on the RCX plays simple tones. Infrared transceivers on the RCX allow it to communicate with computers and other RCXs.

The RCX has plenty of memory for the purposes of FLL, so you shouldn't have any problem downloading programs to control your robots. However, each RCX holds a maximum of five programs, regardless of the amount of memory in use.

The NXT Brick

Figure 9-5 shows the NXT's microcomputer, the NXT Brick. It has four input ports and three output ports (numbered *1* through *4* and *A* through *C*); one more input port than the RCX.

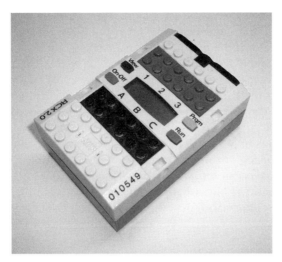

Figure 9-4: An RCX microcomputer

The NXT port connections differ from those on the RCX; they look more like phone jacks.

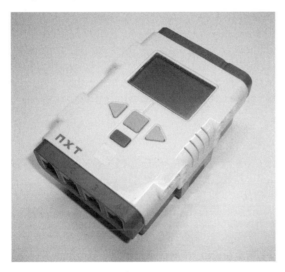

Figure 9-5: An NXT Brick

The display screen on an NXT Brick is significantly more powerful than that of an RCX. It's quite a bit bigger and can display a menu of available programs, pictures, drawings, sound files, and much more. You can even create basic programs right on a Brick, using only this menu!

The NXT Brick has four buttons. A square orange button in the center serves as the On, Enter, and Run button. Use this button to turn the robot on, enter a folder on the menu, and run programs. The rectangular gray button beneath the center button acts as a Clear, Cancel, and Back button.

Use it to go back in the menu, stop a program, or clear an entry. Two gray arrow keys, one pointing left and the other pointing right, allow you to switch between selections in the menu. You can also program the Enter and arrow buttons to act as touch sensors in programs.

The NXT Brick's Bluetooth capability allows it to communicate over a much greater distance than an RCX. For example, using Bluetooth enables you to download a program from your computer to a Brick that's on a different floor!

The speakers in NXT Bricks are also more advanced than those in RCX and can play speech or sound effects in addition to simple tones.

Although users can install more than five programs on an NXT Brick, the Brick's memory is somewhat small; it's easy to go over the limit if you use several large programs and/or sound files.

Batteries

Both NXT and RCX microcomputers can use six AA batteries as a power source. However, NXT Bricks can also use a special lithium rechargeable battery pack made by LEGO. This battery pack snaps into the Brick, and you can use a transformer cable to recharge it even when it's in the Brick.

The addition of this rechargeable battery pack can be a big advantage in FLL, because it enables teams to easily recharge their robots and keep a consistent battery level for each match. This can be important, because robots usually act differently when their battery levels change. It's harder to keep a consistent battery level with AA batteries, because you must exchange them frequently to keep a high charge level when practicing.

Sensors

A robot's sensors are its only source of information about its environment. Although the sensors for RIS and NXT have some similarities, they also have some big differences.

RIS Sensors

The RIS has four sensors, as shown in Figure 9-6. *Touch Sensors* are the simplest and can detect a button on the sensor being pressed, released, or bumped. *Light Sensors* measure the amount of light in their surroundings. The reflected light of an LED on each Light Sensor can be detected by the sensor (this LED is always on when the sensor is used in a program).

Rotation Sensors can be connected to axles and measure how much they rotate. For example, if you were to turn an axle in a Rotation Sensor all the way around, the sensor would measure one rotation. Rotation Sensors are accurate to sixteenths of a rotation.

Finally, *Temperature Sensors* measure the temperature of their surroundings. They are not allowed in FLL.

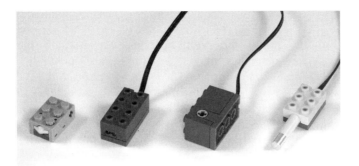

Photo Courtesy of Dave Parker

Figure 9-6: Sensors in the RIS

NXT Sensors

Although NXT robots can work with all RIS sensors using special converter cables, the NXT system includes its own sensors. The redesigned Touch and Light Sensors, as well as new Sound and Ultrasonic Sensors, are shown in Figure 9-7.

Figure 9-7: Sensors in the NXT

The NXT's *Touch Sensor* has the same functions as the RIS sensor, but it's bigger and it looks different. The NXT's *Light Sensor* is larger than its RIS counterpart, but it has almost the same functions, except that you can turn off its LED. The increased size of the NXT sensors is generally considered a disadvantage, because it's harder to fit them into small areas.

Although the NXT system doesn't include separate Rotation Sensors, its motors have built-in Rotation Sensors. These sensors allow you to directly measure the rotations of any NXT motors you use, which can be a big advantage in FLL. Better still, these sensors are accurate to one three-hundred-and-sixtieth of a rotation!

The *Sound Sensor* measures the volume of sound in its surroundings. However, like the RIS Temperature Sensor, the Sound Sensor has no practical use in FLL and isn't allowed on FLL robots.

On the other hand, the NXT's *Ultrasonic Sensor* can be quite useful. It uses sonar to measure the distance between it and the closest object in range. This is useful for both avoiding and finding objects (such as mission models).

Motors

NXT and RIS motors differ both in appearance and performance. RIS motors are significantly smaller than NXT motors. This is generally considered an advantage in FLL, since it allows for smaller robots and lets the motors fit into smaller areas. However, the increased size and weight can come in handy for balancing weight properly or for using motors as the "foundation" of a robot (as discussed in Chapter 10).

RIS motors can rotate faster than NXT motors (when they aren't driving anything). However, NXT motors are much more powerful, enabling you to use gears to make an NXT motor drive something faster and with more power.

Construction Pieces

As mentioned earlier, the RIS and the NXT are designed to work with different construction pieces, one with studs and one without. While there are similarities between the elements (they're all LEGO, after all), there are also some major differences.

RIS Construction Pieces

The RIS was designed to work best with a mix of TECHNIC bricks and pieces such as pegs, axles, and gears. Figure 9-8 shows examples of some of the pieces commonly used in RIS robots.

Figure 9-8: Common construction pieces used in RIS robots

Building with this construction system is somewhat similar to building with regular LEGO pieces, which may make it easier to build robots if you're new to LEGO TECHNIC. However, this system also has a couple of downsides. For example, it can be hard to connect pieces at oblique angles, and it can be difficult to build strong constructions.

NXT Construction Pieces

The NXT construction system uses mostly studless pieces, as shown in Figure 9-9.

Figure 9-9: Common construction pieces used in NXT robots

The lack of brick and brick-like pieces in the NXT construction system, and the fact that its pieces are usually connected by pegs and axles instead of studs, may be hard to get used to. However, once you have a little experience with this method, you may find it easier to build robots with this system. The pieces seem to be better designed for allowing irregular angles in designs and for fitting gears and electrical parts in tight positions. They're also better suited for creating strong designs, and they work well with bracing (as discussed in Chapter 10).

Base Kits

Each system has one or more *base kits*, which you can think of as "starter kits." These kits include a microcomputer, motors, sensors, connector cables, construction pieces, and instructions for building several starter robots. Some base kits also include programming software. Although you aren't required to, you'll probably want to get a base kit instead of trying to get all the pieces you need through accessory kits.

RIS

The latest version of the Robotics Invention System base kit (RIS 2.0) includes the following electrical parts:

- 1 RCX
- 2 motors
- 1 Light Sensor

- 2 Touch Sensors
- 6 connector cables

There are about 700 construction pieces in the kit, consisting of TECHNIC bricks, gears, pegs, several kinds of wheels (including treads), and much more. The kit also includes programming software (RCX Code) and an infrared transceiver for downloading programs to robots.

How well does this set cover the needed pieces for your robot? Well, there's a good chance you'll need another motor for your FLL robot. You might also want some other sensors, such as Rotation Sensors. However, the building pieces will probably be fairly adequate.

NXT

The NXT system has two base kits: the Retail and Education Kits. Both kits sell for about $250.00.

Retail Base Kit

The *Retail Kit* (sold by LEGO Shop at Home at *http://shop.lego.com/* and other retailers) includes the following electrical parts:

- 1 NXT Brick
- 3 motors
- 1 Ultrasonic Sensor
- 1 Light Sensor
- 1 Sound Sensor
- 1 Touch Sensor
- 7 connector cables

This kit has over 500 construction pieces, consisting of studless beams, gears, pegs, axles, and more. Only four wheels are included (while the RIS 2.0 base kit has 20). The kit also includes the NXT-G programming software and a USB cable for downloading programs to the NXT.

Education Kit

The *Education Kit* is sold by LEGO Education (*http://www.legoeducation.com/*). It includes more electrical parts than the Retail Kit: an extra Touch Sensor, three LEGO lamps, a lithium battery, and three converter cables. However, it has fewer construction pieces—about 430 pieces compared to the Retail Kit's 500, and many of them aren't as useful for robots. The Education Kit does not include the NXT-G programming software, which you can buy separately from LEGO Education for about $70.00.

Both the Retail and Education Kits provide a good amount of electrical parts for FLL robots. FLL allows a maximum of three motors on any one robot, and you probably won't need any extra Rotation Sensors, since each motor has one. However, if you get the Education Kit, you may want some extra construction pieces. Also, both kits come with only one or two kinds of wheels, so you may want to get other wheels depending on your robot's needs.

Availability

One important factor to consider when choosing between the RIS and the NXT systems is the availability of each. Since the NXT replaces the RIS (and is not just an addition), it may be harder to purchase an RIS.

As of this writing, LEGO no longer sells RIS base kits, although LEGO Education will sell accessory packs through 2009. However, you may well be able to purchase RIS products through third parties, such as eBay (*http://www.ebay.com/*) and BrickLink (*http://www.bricklink.com/*).

The NXT is readily available from LEGO and other retailers. LEGO also sells many accessories and individual parts, so you shouldn't have any trouble getting the right parts. Also, FLL offers (for a price) an NXT robot set to teams when they register, but not an RIS set.

Choosing a System

Now that we've compared the two systems, let's look at the overall picture. As previously discussed, the RIS seems to have advantages over the NXT in the size of electrical parts and the fairness bonus given to teams that use it. One could also argue that the construction system for the RIS is easier to use, but that's more a matter of taste and experience. The NXT, on the other hand, has a more powerful and versatile microcomputer, a more reliable battery source, a larger and more advanced sensor capability, a stronger motor performance, and a much wider availability.

FAIRNESS BONUS AND PART LIMITS

NOTE   *Future seasons may exclude the fairness bonus. If there is no bonus the season you compete, you may want to move on to the next section.*

Because of the advantages in the NXT system over the RIS, FLL has added *fairness bonuses* to enable teams using the RIS to get extra points. For example, in the Nano Quest season, NXT teams received 20 extra points if they earned points in any 6 missions, while RIS teams only needed points from 3 missions to receive the bonus. In the Power Puzzle season, the fairness bonus basically scaled up RIS teams' scores.

FLL also places different part limits on the two systems. For example, in the Power Puzzle season, one Ultrasonic Sensor was allowed in NXT robots. Since the RIS has no such sensor, RIS robots were allowed to use an extra Touch or Light Sensor instead. Also, NXT motors counted as both motors and Rotation Sensors, since they have these sensors built in.

The advantages of the NXT's battery source and sensors are especially significant. For example, the rechargeable battery pack makes it much easier to keep your robot at a consistent, full battery level, which can greatly enhance its performance. Also, while the RIS Rotation Sensor is only accurate to sixteenths of a rotation, the NXT sensor is accurate to one three-hundred-and-sixtieth of a rotation. This accuracy enables high precision in simple movements, which can be a great advantage in the Robot Game. And since the NXT Rotation Sensors are built into motors, the programming software can use them to keep a robot moving straight automatically by making sure two motors turn at the same speed.

The NXT's stronger motor performance is another great advantage. Even though you might not need the increased power, you can use it to give your robot extra speed to finish missions more quickly.

Another advantage (at least to many teams) pertains to the sturdiness of NXT robots. Because they use pins and axles to connect parts, when compared with RIS-based robots, NXT robots are frequently more lightweight with parts that are less likely to fall off. Many teams who have built RIS robots can recount stories of robots losing pieces on the mat or entire robots shattering after falling just a few feet. Studless building has an inherent strength built into the parts and connection methods that RIS lacks.

One advantage that the RIS does seem to have over the NXT is that it can easily be connected to almost all LEGO brick-like parts. It can be hard to merge TECHNIC beams with regular brick-like pieces, but there's a real advantage in being able to use a wide selection of available brick-like parts (compared with fewer studless parts). Many teams come to FLL with an already large collection of brick-like parts, which can make building RIS robots much simpler.

Remember that the factors discussed in this chapter aren't the only ones to consider. For example, you may have much more experience with one system than the other. Also, the price of each system might influence your decision. Maybe you already have one system and don't want to spend the time and money to switch to the other. You'll want to take factors like these into account besides the ones discussed in this chapter.

SYSTEM TIP

Jonathan Daudelin has had experience using both the RIS and NXT in FLL. The Ocean Odyssey challenge was his team's first year of competing, and they used the RIS. The NXT was released by the next season, and they opted to use it instead. During these two seasons, he grew to prefer the NXT over the RIS, mainly because of the accuracy of its rotation sensors and flexibility of the studless building system. The unlimited number of programs and rechargeable battery pack also proved to offer significant advantages for his team.

10

BASIC BUILDING

Building robots is an art. Using a bunch of pieces to form a working robot is quite an exciting accomplishment! As it turns out, you can use many techniques to refine your skill and proficiency at building LEGO robots. This chapter goes over some of these techniques and gives general tips on building good robots.

NOTE *Since NXT is the newest system used in FLL, and since RIS may be removed in future years, the building techniques discussed here focus on the NXT system. However, many methods may apply to the RIS as well.*

Although we discuss several useful techniques in this chapter, the best way to enhance your skills is simply to spend time building and experimenting with robots. The more you build, the better you become at it!

Using Existing Models

Our first helpful tip has nothing to do with building! If you have a hard time coming up with a design or don't have enough time to create a new one, you might want to simply modify an existing basic design.

For example, you probably want to build a mobile robot for the Robot Game. A *mobile robot* is basically just a chassis that uses two motors and three wheels to move forward, backward, and turn.

You may have already built a basic mobile robot or have instructions for building one, such as the TriBot from the Education or Retail Kits. These basic robots are a great beginning because you can modify them to do what you want. For example, you could add a motor that will flip a lever on a mission model or attach an Ultrasonic Sensor to detect an obstacle.

Our point is that you can take an existing "generic" design, such as a simple chassis, and modify it to suit your needs. Not only does this eliminate some of the building, but it can also help you get started. After working on some of the missions, you might decide to return to that starting design and make changes.

TECH TIP: BASE DESIGNS

Use what you know! Why reinvent the wheel when your team can start a robot design using a preexisting robot? The Internet has readily available designs, such as the JennToo (designed by LEGO MDP/MCP Brian Davis), and you can always use the very popular TriBot design that comes with your NXT-G software tutorials. If time is limited or your team is not very experienced with robot design, start with an existing design and modify it as needed.

Building from a "Foundation"

Building a robot from scratch is trickier than modifying an existing design because you don't have anything to build "off of." Rather, you need to put together a bunch of individual pieces.

When building from scratch, consider using pieces like an NXT Brick or motor as the "foundation" of the robot. For example, to build a rover that uses two motors, each of which is attached to a rear wheel, start building the rover with these motors. Connect pieces to a motor so that it turns a wheel, and then do the same thing with another motor. Next, connect the two motors so there is a wheel on each side of the robot. Finally, add the front wheels, the NXT Brick, and any other parts you want.

Sounds pretty easy, doesn't it? This simple starting method gives you a sense of direction because it helps you see which pieces to use and where to put them.

Building with Modular Design

Modular design consists of connecting multiple *modules*, or components, to form a robot. For example, make two modules, each with two wheels and a motor that turns them. Then make modules for a robotic hand and the NXT Brick, and connect all these modules together to form a rover equipped with a robotic arm.

Using modules when you design can help organize your thought process by breaking up the entire robot into smaller, more manageable components (which can also reduce your work). Another example is building a module for two wheels on one side of the robot and then adding the other two wheels by copying the opposite module. Figure 10-1 shows an example of a module for a tread.

Figure 10-1: A module for a tread

This module consists of a motor that drives a tread. To build a rover that uses treads, simply copy this module on the opposite side, connect the two, and add an NXT Brick.

Bracing Your Design

Have you ever had a design that was not sturdy or strong? If so, you probably could have fixed the problem by bracing the design. Consider the robot in Figure 10-2.

Figure 10-2: A robot that hasn't been braced can have a weak structure.

What's wrong with this robot? It isn't very sturdy, right? The wheels bend away from it, which will harm its performance: The robot might not turn as accurately, and it might move in a slight curve instead of a straight line.

Fix this problem by bracing the wheels, as Figure 10-3 shows.

The beam connecting the two motors is called a *brace*, which keeps the wheel mechanisms from moving toward or away from each other. Bracing your designs makes them much stronger. This in turn can improve their accuracy, which is usually important in the Robot Game.

Figure 10-3: The robot in Figure 10-2 after being braced with beams connecting the motors

Types of Braces

You can brace designs with straight or diagonal braces as Figure 10-4 shows.

Figure 10-4: A straight brace and a diagonal brace

Diagonal braces are often used to stabilize a long object that is only connected at one end. For example, suppose a sensor is attached to a pole that extends several inches above the robot. If this pole is attached in only one place at the bottom, it will probably bend pretty easily. Adding a couple of diagonal braces to connect the top of the pole to the robot makes it much stronger.

When bracing, it's a good idea to use three or more *contact points* (points where the brace connects to the design) to make the brace stronger, as Figure 10-5 shows.

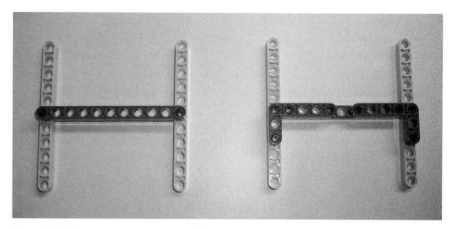

Figure 10-5: Two contact points versus four

The brace with only two contact points doesn't work as well as the one with four, because the beams with two might rotate. Usually, the more contact points you have, the less your connected objects are likely to rotate and the stronger your brace.

Connecting a Brace

You can connect a brace in several different ways. One common way is simply to connect a beam to the desired objects using pegs. However, sometimes this doesn't make a very strong connection, so Figure 10-6 shows two examples of stronger brace connections.

Figure 10-6: Two examples of strong brace connections

Notice how strong these connections are. They won't come off unless you push out those little axles! Also notice that if the other sides of the braces had similar connections, each would have four contact points.

Gears!

Gears are great little pieces that you can use for many different things. They allow motors to turn something that's in one place and point it in a different direction. You can also use gears to adjust the speed and torque with which a motor turns something.

Gear Chains

A *gear chain* refers to multiple gears that are "connected" so their teeth mesh with each other. Figure 10-7 shows a simple gear chain.

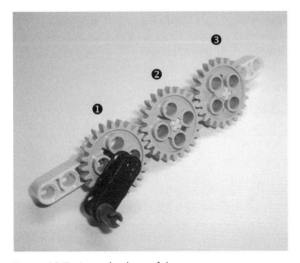

Figure 10-7: A simple chain of three gears

What happens when you turn the handle connected to the first gear? The teeth on its outer rim push the teeth on the second gear, causing that gear to rotate in the opposite direction. The second gear's teeth then mesh with the third gear's teeth, again causing it to rotate in the opposite direction (which is the same direction as the first).

If you use this technique, you don't have to put the motor where the wheel needs to go to power the wheels. Instead, you can put the motor in a convenient place and use gear chains to send power to a wheel some distance from the motor.

Turning Multiple Objects

Use gears to make a motor turn multiple objects. Figure 10-8 shows an example of this. When the middle gear turns, the other two gears rotate in the opposite direction. If you then attach a wheel to each of these gears, you could turn both of them by turning the middle gear.

Figure 10-8: A different configuration of the three gears in Figure 10-7

Using this concept, you could make a four-wheel-drive robot just using two motors: one to power the front and rear wheels on one side and another to power the wheels on the other side.

Joining Gears at Right Angles

Gears don't always have to be connected in the same plane; as long as their teeth mesh well with each other, they should work fine. This means you can even join gears at right angles. Figure 10-9 shows an example.

Figure 10-9: An angled gear chain

As this figure shows, when the larger gear turns, its teeth still mesh with those of the second, smaller gear, causing it to turn. So basically, this concept allows you to turn the direction of motion around 90 degrees.

Gear Ratios

Use multiple gears of different sizes to increase the speed and decrease the *torque* (turning force) delivered by a motor, as Figure 10-10 shows.

Figure 10-10: A gear chain in which the gear ratio is greater than one

As this figure shows, when the large gear turns, the smaller gear turns quite a bit faster. Try it! Each time the large gear makes one complete rotation, the smaller gear has to make several complete rotations to "keep up."

Also notice that it would be hard to turn the large gear with your finger resting on the small gear. When a larger gear turns a smaller gear, the smaller gear turns faster but has less power or torque, since it's more difficult to spin a gear faster. In this kind of chain, the *gear ratio* between the two is greater than one. The gear ratio is equal to the number of teeth on the turning gear divided by the number of teeth on the gear being turned. In the example, the ratio is greater than one, since the larger gear has more teeth than the smaller gear.

Gearing Up

Whenever the gear ratio between two gears is greater than one, the gear being turned increases speed and lessens torque. This is called *gearing up*. In our first example of a gear chain, the gears had the same number of teeth, so their gear ratio was exactly one. When the ratio between two gears equals one, the gear being turned will do so at the same speed and with the same amount of torque as the gear doing the turning.

Gearing Down

Look at Figure 10-11, which shows a different configuration of the two gears in Figure 10-10. When you turn the small gear in this configuration, the large gear turns more slowly. However, it should be easy to turn—even if you rest your finger on the larger gear to slow it —because in this case the gear doing the turning is smaller than the gear being turned, so the ratio is less than one. When this is the case, the gear being turned has less speed but more torque. This is called *gearing down*.

Figure 10-11: A gear chain in which the ratio is less than one

Gearing up or down can be very useful for robots because it allows you to adjust the torque and speed of your motors. For example, to build a robot that is very fast but not very powerful, gear up the motors to turn the wheels faster with less torque. On the other hand, if you need a powerful robot (to push a heavy item, for example), gear the motor down to make the wheels turn more slowly and with more torque.

Using Gears on Motors

Since motors are relatively big and bulky, it can be hard to fit one next to an object you want it to power. It usually works best to place the motor near the object it needs to drive and then use gears to "transfer" the power to the right place. Although the best way to do this will vary, there are a couple of general techniques for using gears with motors.

When constructing a gear chain to transfer a motor's power, it is often useful to have the motor turn a gear attached to a beam. That gear can turn other gears on the beam, which forms a gear chain. Figure 10-12 shows a few ways to do this.

Figure 10-12: Motors turning gears on beams

In each of these examples, two gears transfer the motor's power onto a gear on the beam. Then we can use a gear chain along the beam. In the example on the left, the gear ratio is greater than one. The gear ratio in the example on the right is exactly one.

Sometimes you might want a gear to turn at an angle different from the angle the motor turns. For example, the motor might spin in the vertical plane, while you need the gear to spin in the horizontal plane. Figure 10-13 shows one possible way to change a motor's angle of motion.

Figure 10-13: Changing the angle of motion from a motor

Treads, Ball Casters, and Wheels

Since your robot needs to move around in the Robot Game, you need some sort of mobility system that will enable your robot to move around. Although you could try to build a walking robot, that probably isn't practical. You will most likely want to use some combination of wheels or wheel-like pieces to enable your robot to move.

LEGO has a huge variety of wheel-like pieces, including small wheels, big wheels, treads (also known as *tracks*), and even balls. Figure 10-14 shows some examples.

Figure 10-14: A few different types of LEGO wheels and wheel-like pieces

Treads

A robot using treads usually has one on each side. The robot can turn by rotating the treads in opposite directions or rotating one tread faster than the other. Figure 10-15 shows an example of a robot that uses treads.

Figure 10-15: A robot with treads

Ball Casters

Figure 10-16 shows an example of a *ball caster*.

Figure 10-16: A ball caster

The ball in the ball caster spins freely but can't move out of the container. Ball casters are commonly attached to the front of a robot with two powered wheels in the back to allow it to spin or turn easily, as well as move forward or backward. However, a robot with only a ball caster in front may also easily tip over, since it only has one support in the front.

Wheels

You can use any number of wheels on a robot, but it's most common to use two, three, or four. When you use two wheels they're usually positioned at the back of the robot, one on either side, with a supporting structure in the front (a ball caster or a low-friction LEGO piece). Figure 10-17 shows an example of a robot with two wheels and a supporting structure.

Figure 10-17: A two-wheeled robot with a supporting structure

When you use three wheels, it's common to have two powered wheels at the front or back and a third wheel opposite them attached to a swivel so it can spin around to let the robot turn or move forward or backward. Figure 10-18 shows an example of a wheel on a swivel, called a *caster wheel.*

Figure 10-18: A caster wheel

Four-wheeled Robots

You can use four wheels on a robot in multiple ways. One common way is to have motors power two wheels in the rear of the robot with two wheels in the front that spin freely. The two front wheels purposefully have little traction so they can slide along the ground when the robot turns. This steering method is called *skid steering*.

It's also common to have four powered wheels, called 4WD (four-wheel drive). One motor usually turns two wheels on one side of the robot, while another turns two wheels on the other side. To turn the robot, either the motors rotate in opposite directions or one motor rotates faster than the other. This kind of steering can be difficult, depending on the location of the four wheels. Usually, it works best when the front and rear wheels are very close together. Figure 10-19 shows an example of such a robot.

You can also design a four-wheeled robot to turn using the *rack-and-pinion* method, similar to the steering that cars use. In this design, a motor powers two wheels at one end of the robot, while another robot points the two wheels at the other end in different directions to steer the robot.

Figure 10-19: A 4WD robot

Wheel Size

The size of a wheel affects the speed and power at which the robot moves, much like gears do. If a robot has big wheels, it moves faster than it would with small ones; however, it will also be less powerful. Choose wheels of a size that give you the right balance of speed and power.

The size of wheels also determines how high off the ground the robot sits. If the robot will move over objects, you may want to use bigger wheels to keep the robot from hitting those items.

Traction

The traction of wheels is also an important factor. Having a good amount of traction can make your robot more powerful because the wheels do not slide as easily. Many wheels have rubber tires that give them significantly more traction. Rubber treads are also an excellent source of traction, because a large surface area touches the ground.

Wheel Issues

Watch out for a couple of things when attaching wheels to your robot. As observed earlier, depending on design and bracing, a wheel can bend out (negative camber) from the weight of a robot, which decreases the robot's consistency and accuracy. However, you can usually fix it by shortening the distance of the wheel from its contact point and attaching it to multiple points on the robot. Of course, it's also important to make sure wheels don't rub against a motor or the robot, which could cause a number of problems.

> **TECH TIP: BUSHINGS**
>
> Many teams make the mistake of pushing a rubber wheel onto an NXT motor without including a *bushing*. The bushing provides a small amount of space between the wheel rim and the motor and can help prevent rubbing and friction. Try it—put a rubber wheel on the motor, and push it as far against the motor as possible. When you rotate the wheel, notice how it rubs against part of the motor. Putting a single bushing on the axle will prevent this from occurring.

Building with the Brick

Now for some building tips related to the NXT Brick. The Brick is an important part of a robot, and knowing some techniques for building with it can be very helpful.

Allowing Access

The Brick is the communication link between you and the robot. You use it to tell the robot what to do and when to do it. The Brick can also relay information from the robot to you by its display screen or microphone. Because of this, it may be helpful to build your robot so that you have easy access to the console and display screen.

It's common to attach the Brick to the top of a robot, facing up. This makes it easy to select and run programs on the NXT as well as to see information on the display screen. On the other hand, if you put the NXT Brick on the bottom of a robot or build over it, you might have trouble accessing the buttons and seeing the screen. In FLL, this can put you at a significant disadvantage, since being able to quickly select and run programs is usually important.

It can also be helpful to build your robot so you have easy access to the Brick's batteries. When the batteries run out, you don't want to have to take half the robot apart to change them! There are a few ways to prevent this. You could attach the NXT Brick to the robot in a way that lets you easily remove it, or you could try to build around the battery compartment so that you can take out the batteries without removing the Brick from the robot. If you have the special rechargeable battery pack, you need only allow access to the charging port so that you can plug the charging cable in; the battery doesn't have to be removed from the Brick to be charged.

Finally, it's important to allow access to the cable ports to which you need to connect cables. If you download programs via a USB cable, you also need to leave access to its port. Remember, the ports also need some space around them where the cables will stick out.

Using the Brick as a Counterweight

You've probably noticed that the NXT Brick is heavier than most pieces. Use this to your advantage in some robots by placing the Brick where it will act as a *counterweight*, to balance the weight of something else.

For example, suppose you had a robot that was heavy in the front, and kept falling over because of it. You could attach the NXT Brick to the back of the robot to balance the weight in the front and keep the robot from tipping over.

Attaching the Brick to a Robot

You will sometimes find it easier to build the entire robot except for the NXT Brick, and then connect the Brick wherever is most convenient. This helps you make sure you'll have easy access to the console, display screen, and other important areas.

Although you can connect the Brick in countless ways, Figure 10-20 shows a few common techniques.

Figure 10-20: A few general ways to attach a Brick to a robot

These three demonstrated techniques use the peg holes on the sides and bottom of the NXT Brick. As the middle example shows, the peg holes on the bottom of the Brick are useful for connecting it to beams on top of the robot. When this is impossible, peg holes on the side come in handy. Attach beams to the peg holes, and then connect the beams to the design.

This method of attaching the Brick also enables you to set it at a fixed distance from a part of the frame. As the left and right examples show, you can attach beams to side holes horizontally or vertically, whichever works best for your design.

Figure 10-21 shows another way to connect the Brick using the side holes.

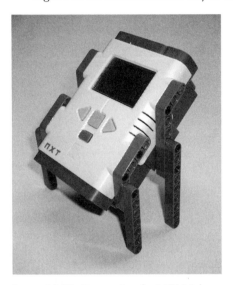

Figure 10-21: Connecting the NXT Brick at an angle

Using this method, you can attach the Brick at an angle, which can be helpful when the robot isn't flat. You can also attach the Brick at an adjustable angle by attaching straight beams to the sides, with taller beams at one end and shorter beams at the other.

Motors

Motors are a robot's source of power; they make it move. Let's go over some techniques you can use when building with motors.

Using Motors as Counterweights

Because motors are relatively heavy, like the NXT Brick, you can use them as counterweights. When you do, though, it's a good idea to point the motor so the back is at the lighter end of the robot, because the heaviest part of the motor is in the back.

Controlling Motors with Move Blocks

When programming in NXT-G, you can use Move blocks (see Chapter 13) to send the robot basic commands to move forward, spin, and so on. Although you can select which motor ports to use, the software uses port B for the right motor and port C for the left by default. You don't have to use the default ports when connecting the motors, but if you don't, you have to change the ports for each Move block you place.

When set to move the robot forward or backward, Move blocks turn the motors in a specific direction. Figure 10-22 shows the direction the motor will spin when set for forward motion.

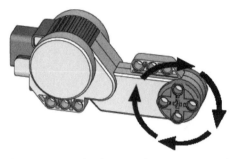

Figure 10-22: The direction the motor spins when set to move "forward"

Of course, depending on how you place motors and use gears with them, this movement may not produce what you consider forward motion. Since it can be frustrating to have a robot moving in the direction opposite how you told it to move, make sure the motors spin the wheels in the direction you want them. For example, if the motors make the robot go backward when set to go forward, consider flipping the motors around to make them spin the right way.

Sensors

Sensors give your robot information about its environment. Although you'll probably spend more time programming sensors than building them into your robot, the following are some helpful tips for effectively integrating them into your design.

Other than the Rotation Sensor, the NXT has three sensors you can use in FLL:

- Ultrasonic Sensor
- Light Sensor
- Touch Sensor

NOTE   *Rotation Sensors are built into motors, so integrating them will not be an issue in most designs.*

Ultrasonic Sensor

In the FLL Robot Game, the Ultrasonic Sensors have three common positions on robots. Probably the most common is to have the sensor pointing straight out from the robot, parallel to the floor. This enables it to detect obstacles near the robot on the mat. Many times an Ultrasonic Sensor is placed at the front of the robot, which allows it to detect obstacles the robot approaches when moving forward.

For example, suppose the robot needs to find a mission model. If an Ultrasonic Sensor points straight out from the front of the robot, you can program the robot to turn until the sensor detects the mission model, at which point the robot should be pointing straight at the model.

You could also position an Ultrasonic Sensor to point straight down toward the ground. Although such a sensor wouldn't be able to detect obstacles around the robot, it could detect objects directly under the sensor, such as the edge of a table. Since the table is a few feet above the floor, as the sensor moves from pointing at the table to pointing at the floor, it would be able to detect the drop.

You could also point a sensor at an angle. For example, if you can't fit a sensor low enough on the robot to detect a short object by pointing the sensor straight ahead, mount it higher up and angle it toward the floor.

NOTE *When you connect an Ultrasonic Sensor to your robot, be sure that nothing interferes with its field of view, including any parts or attachments. The sensor only measures the distance to the closest object, so anything within its range will cause it to ignore objects that are farther away. Occasionally test for interference using the View feature on the NXT Brick to see if the sensor detects part of the robot. How will you know? If, for example, the sensor has a distance of 4 inches but there are no objects near the robot, you'll know to look for interference!*

Touch Sensor

Touch Sensors are simple but quite useful. There are several common uses for Touch Sensors, which can involve different ways to connect them to robots.

Object Detection

Probably the most common use of Touch Sensors is to detect objects. For example, a sensor placed on the outside of a robot can detect an object when a collision with an object, such as a wall, causes its button to be pressed. Figure 10-23 shows an example of this.

The sensor's button is fairly small, and with its limited area, you can see how it could easily miss an object. To increase that area and help to ensure that the robot will detect an object, construct a bumper on the sensor, as Figure 10-24 shows. A *bumper* is a relatively wide part that presses the sensor's button when pressure is applied.

NOTE *Notice that the bumper can't tell where it was pushed, only that it was pushed somewhere.*

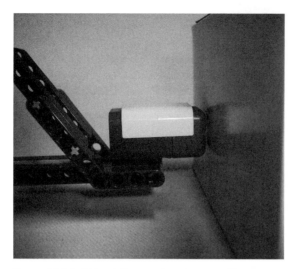

Figure 10-23: A Touch Sensor detecting a wall

Input Detection

You can also use Touch Sensors to allow drivers to give commands to the robot. For example, when a robot returns to Base so the drivers can modify and position it for the next mission, the driver could press a Touch Sensor to tell the robot to start the next mission.

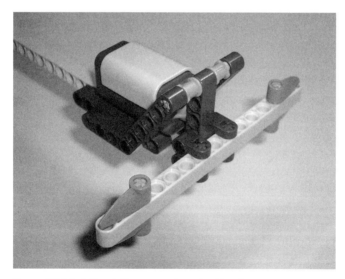

Figure 10-24: A Touch Sensor equipped with a bumper

Light Sensor

The Light Sensor can measure *ambient light* (existing light in surroundings) and *reflected light* (light from the sensor's LED that's reflected by an object). Reflected light can be especially useful for following lines as well as detecting objects.

Detecting Lines on the Mat

To have a Light Sensor detect a line on the mat, point the sensor directly at and close to the floor to allow the light from the sensor's red LED to reflect off the mat back to the sensor. Positioning the sensor close to the floor also eliminates some of the effects of ambient light and enables it to pick up more reflected light.

Minimizing Ambient Light Problems

If you find that ambient light is interfering with a Light Sensor, try limiting the sensor's view with a guard wall. The guard wall will block most of the ambient light, as Figure 10-25 shows.

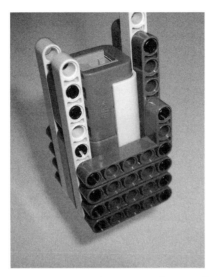

Figure 10-25: A Light Sensor with a guard wall around it to decrease effects of ambient light

NOTE   *You could extend the guard wall all the way down to the mat, but friction between the mat and the wall could interfere with the robot's movements. Also, mats at tournaments have some bumps and irregularities—for example, if the head of a screw underneath is not completely sunk—which could hang up such a wall.*

FOCUSING THE LIGHT SENSOR

One team found a unique way to make a Light Sensor find a color more precisely:

The 'No Limits' challenge in 2005 required the teams to drive down the side of the field to find a white color bus stop. There were three bus stops (two dark, one light), and the position was randomized.

The only way we could solve it consistently was to use the RCX Light Sensors. RCX Light Sensors are inherently myopic and required the robots to run very close to the table edge. When you mix in the imprecision of the system, the robots would frequently run too far away or crash into the bus stops.

The team was quite perplexed by this, and we discussed it over a number of meetings. A few days later I suggested to my daughter that they needed to improve the accuracy. I explained they needed to focus the Light Sensor.

She disappeared and came back with a LEGO magnifying glass. It's about the size of the tip of her pinky. I showed her how to test the optics with a light and a sheet of paper.

About an hour later she came back having built a shroud around the Light Sensor with a perfectly distanced magnifying glass. The result was a whopping 1,200 percent increase in efficiency.

The robot could now run 8 to 10 inches away and still reliably sense the bus stop. What a great lesson learned.

—Marco Ciavolino, TechBrick Robotics, Maryland, USA

Detecting Objects

Although it's uncommon, some teams use Light Sensors to detect objects. If you do so, use guard walls to reduce the ambient light, which becomes more of an issue when the sensor isn't pointing directly at the ground.

Sensing Touch

You can even use the Light Sensor as another Touch Sensor by adding a switch. As Figure 10-26 shows, construct the switch so that a black object covers the sensor when the switch is in one state but removes when the switch is flipped. You would program the Light Sensor to detect both states so the robot can respond accordingly.

Figure 10-26: A switch on a Light Sensor that enables it to be used as a Touch Sensor

Working with Cables

Connector cables are the thick black wires that connect sensors and motors to an NXT Brick. There are three different sizes: 8.5 inches (20 cm), 14.5 inches (35 cm), and 18.5 inches (50 cm). There are a couple techniques for connecting cables efficiently and inconspicuously.

When connecting a sensor or motor with a cable, use the cable that best fits the distance between the electric part and the NXT Brick. This not only helps cut down on slack, but it also prevents the cable from interfering with the robot's performance. A cable with a lot of slack might get stuck in moving components and prevent them from working properly. Cutting down on cable slack also makes the robot look "cleaner."

If you end up with extra cable, a common way to stow it is to loop the cable repeatedly over part of the robot or even around itself. In fact, it's always a good idea to use parts of the robot to hold cables and keep them from interfering with the robot.

Useful Connectors

You will likely find a number of small, basic pieces and constructions useful when building almost any robot. These pieces and constructions help connect parts in tricky positions and at right angles. The following are some of the more common and useful ones.

Hassenpins

When the NXT was released, LEGO made a new part to go with it, as Figure 10-27 shows. Its official name is *TECHNIC Beam with Snaps,* but many people named it *Hassenpin* after Steve Hassenplug, who suggested the design. He designed this part to make it easier to build robots with studless pieces.

Figure 10-27: A Hassenpin

The Hassenpin is incredibly useful, especially when connecting pieces at 90-degree angles. Figure 10-28 shows four beams and Hassenpins making a rectangular construction to use as a frame or the base of a robot.

Figure 10-28: Four beams connected at 90-degree angles by Hassenpins

Another piece similar to the Hassenpin is useful for making a different kind of 90-degree connection, as Figure 10-29 shows.

Figure 10-29: Two beams connected at a right angle

Parallel Connectors

When building a robot, you may find that you need to connect two parallel pieces that do not touch each other. Figure 10-30 shows a useful connector that can help you do this.

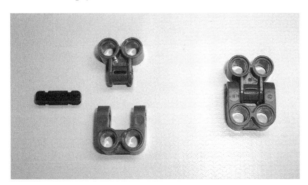

Figure 10-30: Three pieces that can make a useful connector

By using this construction with four pegs, you can connect two pieces parallel to each other quite securely. Figure 10-31 shows an example.

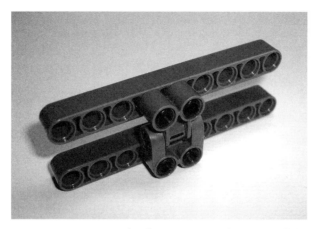

Figure 10-31: An example of connecting two beams parallel to each other

Up Next . . .

In this chapter, we've given several techniques about building robots in general. In the next chapter, we'll discuss some techniques that are more specific to the Robot Game. Also, be sure to check out Chapters 13 and 14 for tips on programming.

11

BUILDING TECHNIQUES FOR THE ROBOT GAME

Chapter 10 discussed some basic building techniques that apply to many different kinds of robots. This chapter explores some more specific techniques that may be useful in the Robot Game. First we discuss some simple but useful techniques, and then at the end of the chapter, we cover a few major techniques related to the design of FLL robots.

Aiming Methods

Due to the high accuracy required in some missions, you may need to position your robot very precisely in Base before sending it on a mission. You can do this in a few ways. One way is to use a sight that is similar to one used to aim a gun. Figure 11-1 shows a robot equipped with a basic sight.

Figure 11-1: A robot with a sight

To use the sight to point the robot at an object beyond Base, hold your eye close to the left beam, and move the robot until both beams line up with that object. You can also position the robot by lining it up with graphics printed at Base, such as the borders. Sometimes the field mat even includes graphics specifically designed for this use. Figure 11-2 shows part of Base on the mat for the Power Puzzle season.

Figure 11-2: Lines around the borders of Base that you can use to help line up the robot

Notice the little line segments around the borders. Teams can align parts of the robot with these lines to help position it correctly. This allows the team to point the robot in a variety of specific directions. However, the features at

Base may not provide the ideal reference points for orienting the robot exactly as desired. In that case, program the robot to turn to the correct position after using the features at Base to align it as close as possible.

Although lining up the robot with graphics can be accurate, it can also be time consuming since it can take awhile to move the robot into just the right position. It may also take extra time and reduce accuracy if you need to turn the robot to a more precise direction after aligning it. Fix these problems by building aiming jigs. *Aiming jigs* are LEGO constructions, separate from the robot, that you can use to help aim the robot. Figure 11-3 shows an example.

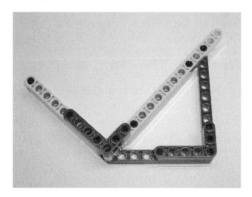

Figure 11-3: An aiming jig

To aim the robot, place the aiming jig in a fixed position at Base, and then push the robot up against it to give it the desired alignment. Figure 11-4 shows an example of this process.

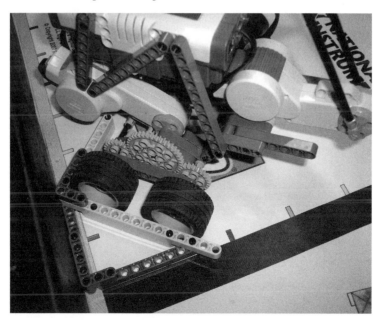

Figure 11-4: Using the aiming jig to position a robot

In the figure, the aiming jig is lined up against the back right corner of Base. The robot is pushed up against the other sides of the jig to arrange it at a specific position and angle. Notice how it ensures that every time you use the aiming jig, the robot always points at the same angle. Although it may take the drivers a while to position the aiming jig, they can do it while the robot is running a mission and then quickly push the robot into position against the jig.

The 4-inch border wall at Base could be considered a permanent aiming jig. You can push your robot up against it to make sure it's straight and at the back of Base. However, be careful about doing this, as the mat may not always be pushed right up against the wall, which could cause your robot's position on the mat (which is what counts) to vary slightly.

Approaches to Handling Table Variety

Although the field mat is ideally flat and completely smooth, real competition tables at tournaments often have bumps and/or aren't completely level. It's a good idea to prepare your robot to handle such imperfections.

Bumps in a table are actually quite common and can cause robots to go off course. The field mat may not be completely smoothed out, or objects, such as a screw that isn't completely screwed into the table, may be under the mat. To minimize the effect of such bumps on your robot's performance, it's helpful if the robot has as little contact with the ground (besides the wheels) as possible. For example, a dozer that slides along the ground on the front of your robot could jolt the robot off course if it encounters something under the mat or goes over a hump. To avoid this, mount the dozer slightly off the ground, or connect it to a hinge so it simply rotates up out of the way when it encounters a bump. Figure 11-5 shows an example of such a dozer.

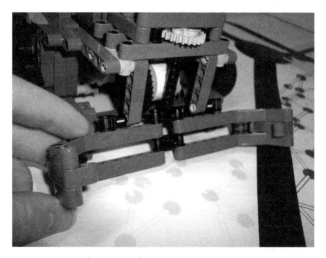

Figure 11-5: A dozer on a hinge

This dozer usually slides along the ground. However, if it encounters a bump in the mat, the hinge allows it to simply lift up and glide over it instead of forcing the whole robot up and possibly off course. Also be careful of pushing large objects across the mat, especially for long distances. This makes the robot susceptible to going off course should it encounter bumps and imperfections in the mat.

A slanted table can also cause your robot's performance to vary. It may help if you give the robot a low center of gravity (by making it reasonably short). Using guide attachments, incorporating sensors, and making the robot frequently return to Base for realignment can also be helpful.

When testing your robot, consider simulating varying table conditions to see how well the robot handles them. For example, you could slip a kernel of rice or a coin under the mat to simulate bumps, or you could prop up one or more corners of the table with books or pieces of wood to simulate a slanted table. If the robot doesn't adequately handle the simulation, experiment with different ways to fix the problem.

Useful Pieces

Even though MINDSTORMS materials usually make up the bulk of pieces used in FLL robots, you can use any LEGO pieces that you want! LEGO has made many interesting and unique pieces throughout the years, and you may want to take advantage of that. For example, one potentially useful piece is the weight brick (*http://www.peeron.com/inv/parts/73090/*); although it isn't very big, it is quite heavy. Potential uses for this piece include balancing weight and holding something in place. Even if you don't have many sets containing unique pieces, you may be able to find them at places such as eBay (*http://www.ebay.com/*) or BrickLink (*http://www.bricklink.com/*), which is an online market only for LEGO pieces and is especially useful for finding single pieces.

Some pieces that may be particularly useful are windup motors. You can use an unlimited number of these pieces, and they can act as temporary motors. For example, you could use a windup motor to make a "minirover" that the robot releases to speed across the table and hit a lever while the robot does something else.

Documentation of Your Design

No matter how hard you try to prevent it, accidents and damage can happen to a robot. It could be as simple as a team member tripping while carrying the robot to another table, causing the robot to explode into a hundred little pieces. You may want to thoroughly document the design so you can recreate the robot if something like this happens. This job would be the responsibility of the Building Backup Manager (discussed in Chapter 6).

One relatively easy way to create documentation is to take photos of your robot. It's a good idea to take plenty of pictures, especially of the detailed components. You may even want to take the robot apart somewhat (making sure you can put it back together afterward!) to photograph in even greater detail. Keep the photos somewhere where they won't get lost.

You can also document your robot by making a *Computer Aided Design (CAD) drawing*, a virtual 3-D copy of your robot on the computer, which you can "take apart" and view from any angle to see how you built the actual robot. Although this may sound complicated, it can actually be quite easy. LEGO makes some special software specifically for doing this called LEGO Digital Designer (LDD). Figure 11-6 shows an example of LDD.

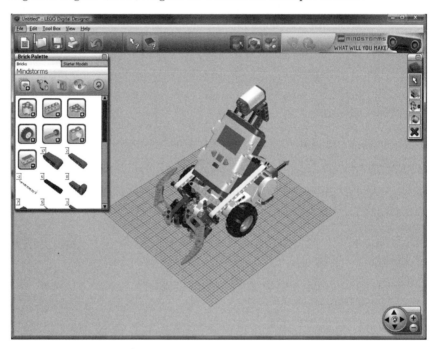

Figure 11-6: LEGO Digital Designer

Download LDD from LEGO's website at *http://ldd.lego.com/*. It's easy to use and can even automatically create building instructions for models you make using it! Although LDD includes many kinds of LEGO bricks and pieces, it doesn't include all of them. However, if your robot uses a piece that LDD doesn't have, you can still document it by taking a photo of the piece and its place in the robot.

LDD has other limitations in areas such as piece placement and gear meshing. MLCAD, another LEGO CAD application, gets rid of many of these limitations, but it is also more advanced and takes longer to learn. Download and learn more about MLCAD at *http://www.lm-software.com/mlcad/*. You can download additional programs to make high-quality building instructions and renderings of models at *http://www.ldraw.org/Article126.html*.

MLCAD has models of almost all LEGO pieces and allows users to make very detailed and high-quality building instructions and design renderings. For example, in Chapter 10, the designs in Figures 10-13 and 10-16 were made using this software. However, its advanced features probably aren't that important for documenting your robot, so you may want to stick with LDD.

Chassis, Attachments, and Bays

Now we get to some major techniques relating to the design of FLL robots. Since the Robot Game has several missions, your robot probably needs several kinds of mechanisms to accomplish them. It might be hard to build one robot with all of those mechanisms, especially with the three-motor limitation. However, the Robot Game is set up so you can add and remove mechanisms during a match! For example, your robot could use a grabber arm to accomplish one or more missions and then return to Base. The drivers could then switch the grabber arm for a long, spinning pole, which it can use to tackle the next mission. In this way, your robot can have a number of attachments and use one or two at a time.

Regularly having the robot return to Base also permits you to realign it, correcting any navigation errors that may have accumulated while attempting previous missions. This method does have a downside, though, in that it can take more time to modify the robot, properly position it, and run the next program. However, you'll probably find that the advantages to this method far outweigh the disadvantages.

It can be helpful to build a robot that uses multiple removable attachments in two or three basic components: a chassis, the attachments, and possibly a bay. Let's discuss each component, starting with the chassis.

Chassis Design

Every robot built for the Robot Game needs mobility to accomplish missions. The main frame of a robot, which contains its mobility system (for example, its wheels), is called the *chassis.* You can consider the chassis the "base" of the robot, because it supports everything that attaches to the robot, such as auxiliary mechanisms. The basic chassis is a good component to build first, because the design of the bay and attachments will probably depend on how you build the chassis.

Size Limitations

The rules for the Robot Game require that all robots fit inside a three-dimensional area called Base. For example, in the Power Puzzle season, Base was 17.75-by-13.25-by-16 inches. All robots must start completely in Base and leave autonomously before accomplishing any missions. For example, a robot would have to completely leave Base before hitting a lever on a mission model. When designing your chassis, consider making it significantly smaller than the size limit imposed by Base (in all dimensions) to allow for the addition of extra mechanisms, such as attachments, which will increase the robot's size in one or more dimensions.

Environment Limitations

Before starting construction on the chassis, look at the field environment in the Robot Game to see how it might affect the design. Figure 11-7 shows the Ocean Odyssey season's mat and mission models.

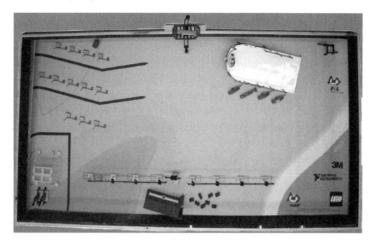

Figure 11-7: The Ocean Odyssey field mat

Apart from the mission models and small bumps and ripples in the mat, notice that the terrain is flat. This means that for this particular competition, you don't need to make an all-terrain chassis; in other words, you probably do not need big wheels equipped with suspension that you might need for an outdoor course. Because the robot doesn't have to go over any small obstacles, it doesn't have to be very high off the ground. You also don't need much traction or power for regular movements, since there are no steep inclines. Therefore, you might want to gear up the motors (as discussed in Chapter 10) to trade some of your power for extra speed.

Mission Model Limitations

The mission models might also affect the chassis design. For example, sometimes a robot needs to push or carry a model across the mat. Since these models are made out of LEGO pieces and usually aren't very big, the robot shouldn't need a lot of power to move them.

However, some missions may require the robot to move onto a model, which can require more power than regular movements. In the Ocean Odyssey challenge shown in Figure 11-7, the mission in the top-right corner required robots to move up a ramp made out of LEGO bricks to get to a submarine model. Having more power in the chassis can also enable the robot to force its way past a mission model if it nicks a corner or strays off course and bumps into it.

Mission models may also restrict the size of robots. For example, the ramp with the submarine restricted the robots' width. If a robot was too wide, it wouldn't fit on the ramp. Watch for tight spaces on the mat that might trap your robot; spend some time using a ruler to take measurements on the mat so your final robot design has the proper dimensions for accurate navigation.

Speed and Power

Chapter 10 discusses using gears and different wheel sizes to adjust the speed and power of your chassis. Of course, the ideal speed and power for the Robot Game will vary from robot to robot, but we can make a few recommendations. Since the Robot Game doesn't usually require much power, and since the NXT motors are more powerful, you may want to gear up the motors to gain extra speed, which can be quite important, considering the two-and-a-half-minute time limit.

Design Procedure

Although you can construct the chassis in many ways, we recommend the following general procedure, which can help your team build the chassis in an effective and organized way.

Before beginning construction, brainstorm with your team, or at least the Building Team. Discuss ideas for possible designs, and then narrow down the suggested designs to a few of the best. Once you have a design short list, split into groups of two or three members (depending on the size of your team and the quantity of building pieces), and have each group build one of the ideas for a chassis. Don't throw away any designs that show potential; the more designs, the better!

Once all groups finish their designs, have another meeting to test and modify them, and then choose the final design (on which you can improve as you test and build).

Temporary Adjustments to Speed and Power

Sometimes a chassis might need a temporary boost of speed or power for one or two missions. For example, one mission in the Power Puzzle season involved two robots (on opposite sides of a table) racing to hit a lever first. In that mission, it was an advantage to have an extra-fast robot, but it might have been a disadvantage during other missions, due to its low power and less accurate movements. Therefore, it would be helpful to make the robot super-fast for just that one mission and slower for the others.

You can temporarily adjust the robot's speed and power in a few ways. One easy way involves simply replacing the current wheels with wheels of a different size. Alternatively, you might stick larger wheels over the existing ones. As Chapter 10 discussed, larger wheels make the robot go faster with less power, while smaller wheels make it go slower with more power.

Adaptable Chassis

To allow for greater adjustments, consider using an *adaptable chassis* like the one shown in Figure 11-8. Although this method is more complicated than simply swapping out wheels, it offers many more options.

Figure 11-8: One side of an adaptable chassis

How It Works

As you can see in the figure, one motor turns three gears on the outside of the chassis (another motor does the same thing for the other side). Above and below each set of three gears are slots for attachments. Figure 11-9 shows one such attachment.

Figure 11-9: An attachment for the adaptable chassis in Figure 11-8

The attachment connects to the chassis using the two beams that extend outward (pegs hold them in). Once connected (see Figure 11-10), the three gears on the chassis mesh with the attachment's gears, allowing the motors to drive the wheels.

Figure 11-10: Connecting the wheel attachment to the adaptable chassis

By attaching another of these attachments to the other side, we get a rover such as the one Figure 11-11 shows.

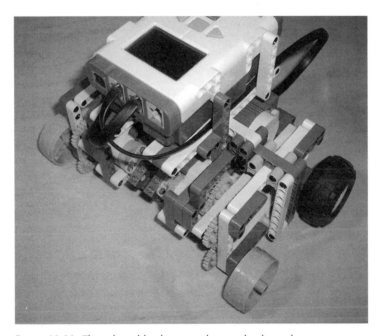

Figure 11-11: The adaptable chassis with two wheel attachments

As you can see, an adaptable chassis like the one in our example allows you to completely change the robot's mobility system simply by switching attachments! The robot could have treads during one mission, wheels during the next mission, and legs during the next; it can use whatever attachments you make for it. This can prove particularly helpful when you need to tackle different types of missions. For example, during a mission that requires the robot to get up on a platform above the ground, you could use an attachment with big wheels or treads. For other missions, you could transform it into less of an "all-terrain" chassis. If the robot needs a lot of speed for one mission, use a geared-up wheel attachment, and then use a slower, more powerful and accurate attachment for other missions.

Challenges

When considering an adaptable chassis, remember that it comes with its own set of challenges. Adaptable chassis can be complicated and somewhat hard to make. Instead of simply driving wheels, the motors need to be able to drive any attachment. You also need a docking system that allows you to securely connect attachments to the chassis; if attachments aren't held firmly, the robot's accuracy and consistency could be affected. Ideally, your docking system should also enable you to quickly connect and remove attachments.

Due to the complexity of an adaptable chassis, its wheels might be farther apart than they would be otherwise. Notice how far apart the wheels are on the robot in Figure 11-6. This can decrease the robot's sturdiness and make the wheels more apt to bend from the robot's weight.

Switching attachments on an adaptable chassis can also take up valuable time during matches. When thinking about using different wheel attachments, consider whether the benefit of that attachment outweighs the extra time involved.

Attachments

Although a chassis can move around, it usually can't perform complicated actions, such as grabbing an object or removing objects from a mission model. To perform these kinds of actions, use attachments. *Attachments* are constructions you can connect to a robot to extend its capabilities by helping it do things like manipulating models, moving models around, and even navigating.

Nonmotorized vs. Motorized Attachments

Many times, it's possible to use a nonmotorized attachment to enable your robot to accomplish a mission. These attachments are simple tools the chassis can move to manipulate models. For example, a mission might require a robot to push an object off a high mission model. To solve this mission, the robot could use an attachment that simply extends above the chassis to hit the object.

Sometimes, a nonmotorized attachment just won't cut it. For example, you might need to drop an object into a specific place on a mission model. To accomplish this, you might use a motorized attachment similar to the one in Figure 11-12.

This attachment has a "bucket" that a motor can raise or lower, which enables the robot to carry a small object and drop it at the correct location.

Figure 11-12: A motorized attachment

Temporary vs. Permanent Attachments

Many of the attachments you use will probably be temporary and won't stay on the robot for the whole match. However, some attachments will be of such general use that you will want them to be permanent features of the robot. Connect these attachments in such a way that they won't interfere when connecting other attachments. For example, you might want to have a dozer on the front of the robot to push mission models around the mat. Simply connect this attachment to the front of the chassis, and keep it on for the entire match. Figure 11-13 shows an example of such an attachment.

Figure 11-13: A permanent "dozer" attachment

Multipurpose Attachments

Sometimes you can save time performing missions by making multipurpose attachments, which the robot uses to accomplish multiple missions. For example, you could make an attachment that has both a hook, to grab a model in one mission, and a stick, to push an object off a high platform in another mission. Even if the robot returns to Base for repositioning, you would not need to change attachments. Of course, the robot might not need to return to Base between missions, which would save even more time.

An attachment might not always work for multiple missions as it is, but could work if you slightly modify it between missions. Modifying an attachment instead of replacing it saves time as well as extra pieces. Figure 11-14 shows the "bucket" attachment in Figure 11-12 after modifying it to work on another mission.

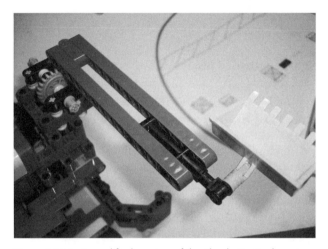

Figure 11-14: A modified version of the "bucket" attachment in Figure 11-12

Instead of removing the first attachment and replacing it with the new one, simply remove the old bucket from the attachment and attach the new bucket construction. This can save time and won't require two copies of the rest of the attachment.

Guide Attachments

At the beginning of this section, we mentioned that attachments can help the robot navigate. How can they do that? Suppose that after traveling to a mission model, your robot needs to be at a specific place to work properly, but there may be some variability in exactly where the robot arrives at the model. This is where a guide attachment can help the robot realign itself. Figure 11-15 shows an example of a funnel attachment that accomplishes this very well.

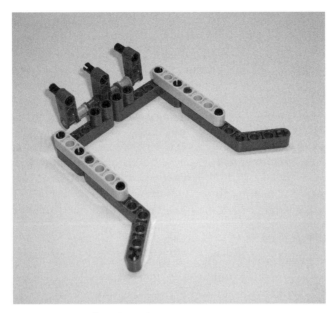

Figure 11-15: A funnel attachment

The two angled beams at the ends of the attachment help align the robot in the right position, similar to the way a real funnel guides liquid to the right place. Figure 11-16 shows how this works.

Figure 11-16: A robot using a funnel attachment to realign itself

Even though the robot is off course initially, the funnel attachment forces it to curve to the left when it tries to move forward. This can be a great advantage because the robot can automatically realign itself without even knowing what it's doing (as opposed to using sensors to try to correct its position).

Guide attachments can take many different sizes and shapes, and their designs will depend on each scenario. Basically, you want to make constructions that take advantage of stationary objects, such as a mission model or wall, to force the robot into the correct position. An object used to guide the robot could be as simple as a wall at the back of the robot that enables it to straighten itself out by backing into the table border. If you consistently have trouble getting the robot to the right place, look for opportunities to use guide attachments.

Size Limitation

Attachments may sometimes extend significantly past the chassis, so it's important to make sure they don't cause the robot to exceed the size limitations. Just as with building the chassis, keep the size limitations in mind when building attachments. Pay special attention to the height limitation, since it is easy to exceed without using a tape measure. When there's a significant advantage to using a larger attachment, you may be able to make an attachment that extends or "unfolds" after the robot leaves Base.

Design Procedure

The design of your attachments depends largely on their respective missions. Therefore, you may want to build the attachment for each mission as you work on it. In other words, build the first attachment when you work on the first mission and the second attachment when you work on the second mission. After designing an attachment for one mission, look at later missions to see if you can use it as a multipurpose attachment, and look at previous missions to see if you can reuse an existing attachment (perhaps by modifying another attachment).

Attachments may also affect the order in which you run missions. For example, if you notice that the robot could use the same attachment for multiple missions, you may choose to run those missions in a row to avoid making unnecessary attachment switches.

Bays

It's often helpful to have a single place on a robot where you connect each attachment. This makes attachments easier to build since you don't have to figure out how to connect each one; you can simply use the bay. It also makes the robot more organized. A simple bay, as shown in Figure 11-17, is a platform that allows you to easily and securely connect attachments to the robot.

This bay would be built onto the rest of the robot, and attachments would snap on and off the four black pegs shown in the figure.

Motorized Bays

A simple bay like the one in Figure 11-17 usually isn't sufficient, since it doesn't have a motor with which to drive attachments. Since you can't bring more than three motors to the competition table, you probably won't be able to have one motor for each attachment, but you can share motors among attachments.

Figure 11-17: An example of a simple bay

One way to share motors when switching attachments is to remove the motor from the first attachment, connect it to the second, and then connect the second attachment to the bay. However, this can eat up a lot of valuable time, so it's usually much more efficient to have a motorized bay. Just like a simple bay, a *motorized bay* also has a motor that can drive connected attachments. Sometimes, the bay may consist of only a motor! In that case, you would simply connect attachments directly to the motor, such as the example that Figure 11-18 shows.

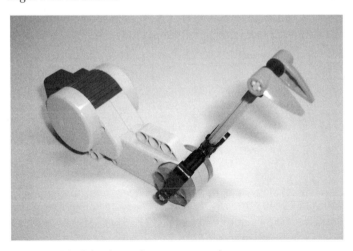

Figure 11-18: A hook attachment connected to a motor

A More Complex Motorized Bay

Depending on the size and complexity of the attachments, simply using a motor for a bay might not work so well. For example, having only one connection to the motor (the axle) might not be strong enough for larger attachments. Figure 11-19 shows an example of a more complex bay that can handle a wider variety of attachments.

Figure 11-19: A more complicated motorized bay

When the motor spins, the black knob gear on top rotates. You can simply connect attachments that don't need to be motorized to the four pegs and connect motorized attachments to the gear on the bay, as Figure 11-20 shows.

Figure 11-20: An attachment (left) connected to a motorized bay (right)

When the motor in the bay turns, it turns the gear on the attachment, which causes the hook to move up or down.

Bay Design

While you could build the bay after completing the chassis, it's often easier to build the two at the same time. This method enables you to *mold* the chassis around the bay instead of trying to fit the bay into an existing chassis design.

NOTE *If you opt to build the bay after you build the chassis, lead a brainstorming session, and then have a few members build the bay. Since it usually doesn't work well to have several members try to build the same thing, other members who want to help could provide feedback and suggestions on the builders' work.*

Web Resources

Although this chapter discusses several building techniques for the Robot Game, you can learn many others from resources such as the Internet. If you have specific questions, look for LEGO robotics–related email groups or forums. If you're looking for general techniques, try tutorial websites or books. Also, browse through online robot galleries (such as NXTLog) to get ideas or inspiration. The appendix lists several helpful websites you may want to check out.

12

SENSORS

When it comes to building and programming LEGO MINDSTORMS robots for an FLL competition, one of the biggest benefits available to you is the use of sensors. Unfortunately, many teams choose not to use sensors in the competition. While there are probably many reasons for not using them, we found that the most common reason is that teams are simply unfamiliar with how to properly program a robot to use sensors.

We hope to change that with this chapter. The sensors that are allowed in competition can be extremely useful not only because they help reduce trial-and-error runs during practice but also because they give your robot flexibility, as we explain further.

NOTE   *As of this writing, teams may still use the RCX in competition. Although we specifically cover the NXT sensors in this chapter, much of the discussion applies to the RCX sensors as well.*

FLL-Approved Electronics

While the NXT system can use a variety of sensors, FLL limits the type and quantity of sensors and electronics used in competition. For example, during the Power Puzzle season, Rule 7 limited each robot to the following electronic components:

- 1 NXT Controller (Brick)
- 3 NXT motors (with built-in Rotation Sensor)
- 2 Touch Sensors
- 2 Light Sensors
- 1 lamp (from the Robotics Invention System)
- Rotation Sensors (3 minus the number of NXT motors used)
- 1 Ultrasonic Sensor
- Wires and converter cables as required

NOTE *FLL may change the quantity and/or type of electronics at any time. For example, as of this writing, third-party sensors (such as the HiTechnic Compass Sensor) are not allowed in competition. Always refer to the latest version of the current FLL season's rules for the approved electronics.*

In this chapter, we cover the function of some of these sensors and how to use them in competition (you will learn how to program sensors in Chapters 14).

NXT Controller

While the NXT Controller (also known as the Brick) isn't technically a sensor, it has some unique capabilities that are useful in competition. First, its three built-in timers prove very useful for keeping track of time. When a mission ends, you can send the amount of time the mission took to the Brick's LCD screen, which means anyone assigned to time the robots during practice can leave his stopwatch at home. (Chapter 14 discusses how to add timer functionality to your programs.)

Another nice feature of the Brick is the ability to provide live sensor readings on the LCD screen. This allows you, for example, to use the Light Sensor to learn the live value for the light level in a room or use the Ultrasonic Sensor to discover the distance between the robot and a wall.

NOTE *Thanks to John Hansen for creating the NeXT Screen tool, which we used for the following images of the LCD screen. Visit* http://bricxcc.sourceforge.net/utilities.html.

To obtain live sensor readings, refer to the following:

1. Connect a sensor or motor to the robot. Note the port number where you attached the device. (For this example, we attach an Ultrasonic Sensor and measure its distance in inches from a wall or object.)

2. Turn on the Brick, and use the Left/Right buttons to select **View**; then press **Enter** (Figure 12-1).

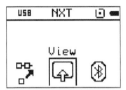

Figure 12-1: Scroll to find and select the View option.

3. Select the **Ultrasonic inch** option as Figure 12-2 shows and press **Enter** (choose **Ultrasonic cm** to display the result in centimeters).

Figure 12-2: Select the Ultrasonic inch option.

4. Select the port number for the Ultrasonic Sensor (**Port 4** in this case) and press **Enter** (Figure 12-3).

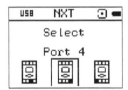

Figure 12-3: Select Port 4 for the Ultrasonic Sensor port.

5. The LCD screen should display the distance from the Ultrasonic Sensor to the wall/object; Figure 12-4 shows a sample reading of 9 inches.

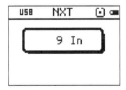

Figure 12-4: Distance from Ultrasonic Sensor is displayed.

6. If you select an incorrect Port number in step 4, you'll see the screen that Figure 12-5 shows. If you get this error, press the dark gray **Cancel** button, and then return to the screen that Figure 12-2 shows to select the proper sensor measurement and port number.

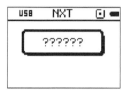

Figure 12-5: Selecting an incorrect Port number

Available Sensor Measurement Options

The View option built into the Brick is an extremely useful feature, with many more options than simply obtaining the number of degrees or rotations that a motor spins. You can also select the following types of measurements using the View option:

Reflected Light   The NXT measures reflected light using the Light Sensor, so select this option to obtain the value for this reading.

Ambient Light   The Light Sensor also measures ambient light; select this option to obtain the value for ambient light.

Light Sensor (RIS version)   Use this to measure the light value with an RCX Light Sensor.

Rotation (RIS Rotation Sensor)   This option displays the rotation value of the RIS Rotation Sensor.

Motor Rotations (NXT)   This option measures the number of rotations the motor has turned.

Motor Degrees (NXT)   Select this option to measure the number of degrees the motor has turned.

Touch (NXT)   This option displays *1* when the sensor's button is pressed and *0* when the button is not pressed.

Ultrasonic inch (NXT)   Use this option to view the distance (in inches) between the sensor and a nearby object.

Ultrasonic cm (NXT)   This option displays the distance (in centimeters) between the sensor and a nearby object.

In Chapter 14, we discuss how to use these values in programs to give your robot the ability to make decisions based on external factors such as lighting, distance from objects, and distance traveled.

The Sleep Feature

While we're on the subject of the NXT Brick, we bring your attention to a feature that some call an annoyance and that isn't often talked about. The NXT Brick is configured to turn itself off automatically after a certain period of inactivity; the default is 10 minutes. While it's a good idea to have your Brick go to sleep between practice runs to save battery power, the last thing you want during the competition is for your robot to turn itself off as it waits for the game to start.

During competition, set the Sleep mode to 30 minutes, 60 minutes, or Never by following these steps:

1. Turn on the NXT Brick.

2. Select **Settings** (Figure 12-6) and press **Enter.**

Figure 12-6: Select the
Settings option.

3. Select **Sleep** and press **Enter** (Figure 12-7).

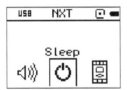

Figure 12-7: Select Sleep.

4. Use the Left/Right buttons to cycle through the options, and press **Enter** when the desired setting appears on the screen. (In Figure 12-8, we selected **Never** so the robot only shuts down if a user turns it off or the batteries die.)

Figure 12-8: Select a
Sleep time option.

NXT Motors and the Built-in Rotation Sensor

The RIS can measure the rotations of a motor using a Rotation Sensor. On the NXT, the Rotation Sensors are built into the motors, and you can program each motor to rotate (forward or backward) a specific number of degrees or rotations, allowing you to fine-tune your robot's movements.

Use the Brick's onscreen View option to obtain the number of degrees or rotations required for a robot to make a specific turn. The value obtained is often extremely accurate, but at other times, it will require some testing and fine tuning.

For example, Figure 12-9 shows how a robot might pivot 90 degrees to its right.

NOTE   *The LCD screen is typically considered the robot's "face," and left/right is based on the direction the robot faces. Left and right are typically referenced from the robot's perspective.*

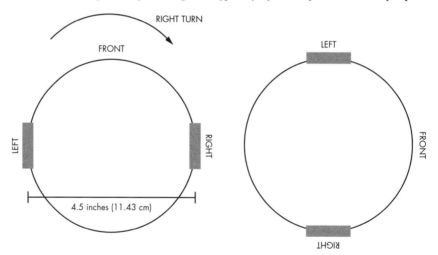

Figure 12-9: How a robot rotates in place

In Figure 12-9, we want the robot to rotate clockwise. The wheel on the right will not move, and the wheel on the left will rotate so the robot makes a right turn. Test this for yourself by building a simple robot such as a TriBot (two motorized wheels and one using a caster or other type of wheel).

Before proceeding with your measurement, connect all the motors that will move the robot. When using two or more motors, be sure to connect them all before taking your measurement. Once everything is in place, use the following steps to obtain an accurate reading of a motor's spin in degrees:

1. Turn on the NXT Brick.
2. Select **View** (see Figure 12-1) and press **Enter**.
3. Select **Motor degrees** as Figure 12-10 shows and press **Enter.**

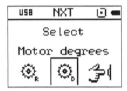

Figure 12-10: Select the
Motor degrees option.

4. To pivot the robot to the right, turn motor B forward (preventing motor C from rotating), and observe the value on the LCD screen that represents the number of degrees that motor B rotates. To do so, choose **Port B** (Figure 12-11) and then press **Enter.**

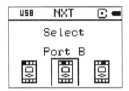

Figure 12-11: Select the
Port B option.

5. Once you select port B, the value for motor B is displayed on the screen; if the robot has not been moved, a value of 0 degrees should appear, as Figure 12-12 shows.

NOTE   *If the value displayed is positive, this indicates that the motor is rotating in a forward direction. If the value is negative, it indicates that the motor is rotating backward. Take this into consideration when programming the motors for forward/backward direction.*

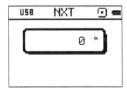

Figure 12-12: An initial
value of 0 for a robot that
hasn't rotated

6. Slowly rotate the robot to the right, and watch the value on the LCD. The final value displayed tells you the number of degrees you need to turn your robot so it pivots. This value depends on the wheel sizes you use as well as how exactly you measured the turn. Record the value so you can refer to it later when programming your robot to make a 90-degree right turn. Note that if the value for a forward B motor movement is positive (+) 183 degrees, then the movement for motor B to spin in the opposite direction (and return to its starting position) will be negative (−) 183 degrees. After motor B rotates and the robot turns 90 degrees to the right, the LCD screen displays a new value (Figure 12-13).

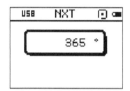

Figure 12-13: A value of 365 was obtained for our robot turning 90 degrees.

Follow the above steps to obtain values in units of rotation, but be warned that the readings will appear as integers (*1, 2, 3,* and so on); the NXT can only display integer values of rotations and not partial rotations. You will never see a displayed result such as *3.2* or *5.9* rotations; instead, the NXT rounds the values to *3* and *6*, respectively.

If you wish to use rotations instead of degrees, simply choose Motor rotations in place of Motor degrees as the measurement option.

Rotation Confusion

A confusing situation that often comes up is this: "If I need my robot to rotate 90 degrees to the right, why can't I simply program the robot to rotate 90 degrees?" In our previous example, we obtained a value of 365 degrees, so why does it take a value of 365 degrees and not 90 degrees to rotate to the right?"

Take a look at Figure 12-9 again. To calculate the rotation distance for motor B, you need values for a few items first, including the tire diameter, tire circumference, and the distance between the tires (the radius, measured from the center of the right tire to center of the left tire). For our example, we use the standard rubber tires that come with the NXT Retail and Education Kits. We also use the TriBot model (which you can build using either kit) to measure the distance between the tires. We need the following values:

- Tire diameter = 2.25 inches
- Tire circumference = approximately 7 inches
- Distance between the tires = 4.5 inches

If we rotate our robot to the right (or left), we need to determine the total distance the rotating wheel attached to motor B will turn when the movement is completed, as Figure 12-14 shows.

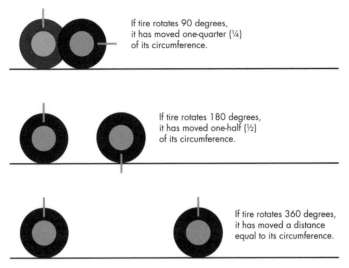

If tire rotates 90 degrees, it has moved one-quarter (¼) of its circumference.

If tire rotates 180 degrees, it has moved one-half (½) of its circumference.

If tire rotates 360 degrees, it has moved a distance equal to its circumference.

Figure 12-14: Motor B's tire will move a specific number of inches to make the right turn.

Examining Figure 12-14, we can determine that if motor B rotates 90 degrees, the tire will rotate 25 percent of its circumference, because 90 degrees is one-quarter or 25 percent of the total number of degrees in a circle. We multiply 0.25 (25 percent) by the tire's circumference (7 inches) to obtain a result of 1.75 inches. This means that if motor B rotates 90 degrees, the tire mounted on motor B moves a total distance of 1.75 inches.

But a tire rotating 1.75 inches isn't enough for the robot to rotate the required distance; if you try it using the TriBot, you'll see that the robot doesn't make a complete 90-degree turn to the right. Motor B must rotate some more to make the actual 90-degree turn.

Hopefully, you can see where the confusion arises: While motor B rotated 90 degrees, the wheel did not make the correct number of rotations to actually move the robot the distance that the arc requires. The number of rotations the robot turns doesn't equal the number of rotations the motor spins; when the motor rotates 360 degrees, the wheel has actually moved 7 inches (the tire's circumference) on a flat plane (Figure 12-14).

Now that we understand there's not a one-to-one correspondence between the motor's rotation and the robot's turning angle, let's calculate the amount that the wheel needs to rotate (in degrees) to turn the robot the required 90 degrees:

1. Determine the complete distance, or the circumference, the robot will travel if it rotates motor B to turn in a complete circle (360 degrees), returning to its starting point. The formula for this is simply pi (3.14) multiplied by the diameter of the circle in which the robot turns, where

the diameter is 2 times the radius. The radius of the circle is the distance between the centers of the two tires (the width of the robot), which is 4.5 inches. Therefore, motor B's total movement would be 3.14 × 9 inches (9 inches is the diameter: 2 × 4.5 inches), which is 28.26 inches.

2. For the robot to make a complete rotation (rotating only motor B), the wheel on motor B needs to travel 28.26 inches. However, we're only rotating our robot 90 degrees, or one-fourth (0.25) of a complete rotation. Therefore, the wheel on motor B should travel 0.25 × 28.26 (or 7.065) inches.

3. Now divide 7.065 inches by 7 inches (the circumference of one wheel) to get a value of 1.009. This is the number of rotations the wheel has to make to complete the pivot turn. Finally, convert to degrees by multiplying 1.009 by 360 degrees. The result, 363.24 degrees, is very close to the actual measured value we obtained using the View option that Figure 12-13 shows.

4. The numbers are almost identical (365 versus 363.24); the point we want to demonstrate is that you can calculate distances fairly accurately before you ever attempt to program the values into your program. Every robot is different, so you'll quickly find that you will often have to make some minor tweaks to the values for degrees and rotations to get the robot to perform the exact movements you desire. In this example, if we had programmed our robot to rotate 363 degrees and run the program, the robot would have come close, but not exact, to the actual 90-degree-arc turn we desire. With tweaking, we would find that a value of 365 would get us the desired rotation.

NOTE   *When programming your robot, you always need to fine-tune calculated values through testing to get the actual value used by your robot. You can also program the motors to rotate for a specific period of time (in seconds), but this is generally considered not as reliable for guaranteeing that a motor will rotate a specific number of degrees or rotations.*

Touch Sensor

The NXT Touch Sensor (shown in Figure 12-15) is a device that can provide your robot with simple decision-making abilities (for example, *if the button is pressed, turn right* and *if the button isn't pressed, continue in a straight line*). This sensor has three settings: Pressed, Released, and Bumped. Here's how each of those settings works.

Pressed

When the Touch Sensor button is pressed and held down, we consider it *Pressed*. The robot's program generally interprets *Pressed* as a collision with an object or wall. While this is the typical use for the Touch Sensor, you can also use the sensor to prevent your robot from going over the edge of a table.

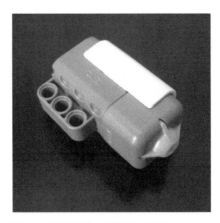

Figure 12-15: The NXT Touch Sensor

For example, if you mount the Touch Sensor pointing downward, with the button always pressed by the pressure exerted against the tabletop, you could use the sensor to detect when the robot reaches the table's edge. When the Touch Sensor reaches the table's edge and the button is released (when the tabletop disappears, the sensor button is no longer pressing against it), you can program the robot to react to this changed condition. If the sensor is mounted away from the body of the robot and its wheels (possibly on a long beam in front of the robot—see Figure 12-16), you can use it to give the robot time to stop and adjust its course.

Figure 12-16: Using a Touch Sensor to detect the edge of a table

Released

When the Touch Sensor is mounted, for example, on the front, rear, or sides of the robot, it can detect collisions with objects (or walls) from the side on which the sensor is mounted. Add clever design to a robot using beams and other TECHNIC components to create a bumper that expands the sensitivity of the Touch Sensor by increasing the surface area that triggers the sensor (demonstrated in Figure 10-22 in Chapter 10). Notice that the robot can detect collisions from a much larger area than the size of the button.

Bumped

The Touch Sensor is considered *Bumped* when its button is pressed and released in less than one second.

You can use the Bumped state as a start button for the robot. For example, we can start the robot's program by pressing **Enter** but have the robot wait to begin its actions until the Touch Sensor is Bumped. (To do this, simply place a Wait block at the beginning of your program, and configure it to wait until the Touch Sensor is Bumped. Once that happens, the rest of the program will begin to execute.)

NOTE *When using the Touch Sensor's Bumped condition, be sure to place the sensor so it isn't accidentally bumped when the robot begins to move.*

Figure 12-17 shows an example of a robot with a Touch Sensor mounted for easy access.

NOTE *Chapter 14 provides an example for programming the Touch Sensor.*

Figure 12-17: The Touch Sensor used as a start button

Light Sensor

The FLL mat used in the Robot Table portion of the competition is very colorful, with a mixture of graphics and text, each of which you can use as landmarks. Use your Light Sensor to distinguish between these landmarks and control your robot's movements much more accurately.

NOTE *The mat will change year to year, so teams should always take a good look at the colors, lines, graphics, and other items on the mat and determine if any will make good landmarks, waypoints, or lines to follow using the Light Sensor.*

Figure 2-1 in Chapter 2 shows the mat used in the Power Puzzle competition. The image is not in color, but it helps to point out a few important items:

- Black lines (representing roads) crisscross the mat at various locations.
- A large blue mass indicates water, and brown graphics represent the boat dock.
- Red squares represent houses.

The NXT Light Sensor (Figure 12-18) can distinguish between most colors by converting them to grayscale and comparing the reading to built-in values.

Figure 12-18: The NXT Light Sensor

Chapter 14 provides a sample robot program that uses the Light Sensor. For now, the following are some examples of how you might use the Light Sensor and landmarks on the mat to better control your robot:

- Program the robot to reverse direction for one rotation when the Light Sensor detects the color blue.
- Program the robot to follow a black line (a road) to ensure movement in a straight line.
- Program the robot to use color to distinguish between "good" objects that must be returned to Base and "bad" objects that should stay in place.

NOTE *Besides detecting color, the Light Sensor can also detect differences in the level of lighting. This ability isn't used very often in FLL competition (the robot is typically operating in a well-lit room with no dark areas on the mat). However, an interesting trick takes advantage of the Light Sensor's ability to detect changes in light readings, which essentially converts a Light Sensor to a third Touch Sensor (as demonstrated in Chapter 10).*

Ultrasonic Sensor

The Ultrasonic Sensor (shown in Figure 12-19) is one of the more powerful sensors available. It gives a robot the ability to measure distances from objects and walls and to detect obstacles. And, unlike the Touch Sensor, which requires the robot to collide with an object or wall, the Ultrasonic Sensor allows a robot to avoid collisions altogether.

NOTE *Chapter 14 covers programming a robot to use the Ultrasonic Sensor.*

Figure 12-19: The NXT Ultrasonic Sensor

The Ultrasonic Sensor works by sending out a sound wave that bounces off objects and walls and returns to the sensor, like radar or sonar. The NXT Brick measures the time that elapses while sending out the signal, bouncing it off an object, and the sensor receiving it again. This detection ability is surprisingly accurate; perform some tests of your own to verify that the sensor correctly measures the distances (in inches or centimeters).

With one Ultrasonic Sensor, this detection ability works well, but more than one sensor in an immediate area can create pandemonium if the signals interfere with each other. If another team is also using an Ultrasonic Sensor on its robot, the signals are indistinguishable from one another, so your sensor might get inaccurate readings if it detects the other team's signal instead of your own.

One way to overcome this problem is by using the 4-inch-high walls of the Robot Table. By mounting the Ultrasonic Sensor low on the robot so its sound wave bounces off the wall, a sound wave sent out by your Ultrasonic

Sensor will (hopefully!) be blocked and bounce off the wall, not interfering with the other team's robot (assuming the other team is also using the Ultrasonic Sensor).

Sensors vs. No Sensors

Not all FLL teams use sensors when fielding their robots; there are pluses and minuses to using sensors and ways to avoid their use altogether. Many teams continue to program their robots using only the simple NXT-G Move block. In fact, through careful testing and measurement, teams can use the Move block to program their robots to move very specific distances (to one degree or one one-thousandth of a rotation). This type of pre-programmed movement is often referred to as *dead reckoning* and *point-and-shoot* movement.

Problems do come with using only Move blocks, however. Some common ones are related to the robot's inability to react to changes in its environment. Consider the following potential problems:

- Extremely small differences in width and depth of competition tables can result in a robot moving too far or too little over long distances. For example, if the table has been constructed one-eighth of an inch deeper than regulation and the mat is flush against the back wall, you need to program your robot to travel one-eighth of an inch farther than normal if its starting position doesn't change.

- If a robot rolls over a small object on the table, it can alter its course drastically, resulting in a table rescue and a loss of points. When you don't use a sensor, your robot cannot respond to changed conditions.

- Multiple activities on the mat require very precise movements. If the robot is off by just a little bit during one mission, that error can trickle down and affect subsequent missions before the robot returns to Base.

- The battery level of a robot can affect its movements. For example, a robot at 70 percent power might move differently than one at 100 percent power.

- Robots that do not use sensors may not be able to correct their course easily. When using dead reckoning, be sure to position a robot very accurately in Base before attempting a mission.

This is not to say that sensors don't have their own problems. For example, as we already mentioned, the Ultrasonic Sensor can return incorrect readings if a second one is nearby, but consider a few other things when using sensors:

- Light Sensors can sometimes be extremely finicky when it comes to detecting colors, let alone the difference between black and white. Be sure to test your Light Sensors in a variety of lighting conditions.

- If a Light Sensor is close to the floor, bumps on a field mat can hamper its ability to correctly detect lines or colors, since they may affect the way that light is reflected.

- For Touch Sensors to distinguish between a Pressed state and a Bumped state, the pressing action must last longer than one second. This means the sensor's button must be solidly pressed and not just "brushed" by a passing object if it is to be considered in a Pressed state. If the button is quickly brushed by a nearby object, the Touch Sensor will treat this collision as a Bumped action, since the press-and-release took less than one second. Program your robot accordingly.

- The Rotation Sensors (built into the NXT motors) work very well when the motors are rolling on a completely flat surface. You can measure distances accurately, but if one or more wheels go over a small bump or object, the programmed distance hasn't changed but the distance traveled on the surface has changed (although it is a very small change) as seen in Figure 12-20.

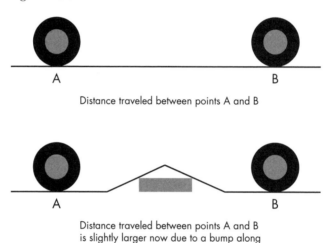

Distance traveled between points A and B

Distance traveled between points A and B is slightly larger now due to a bump along the path, possibly caused by an object or an imperfection on the floor.

Figure 12-20: A small bump on the table can have a big impact on your robot's movement.

Sensing a Trend

Sensors have the potential to be a robot's most valuable secret weapon. When integrated and properly programmed, your robot moves from a simple "point-and-shoot" device to a responsive, "intelligent" teammate. When every obstacle and imperfection on the mat becomes a factor in the success or failure of a robot programmed with only Move blocks, sensors can provide your robot with a real advantage in navigating its way through missions. Give serious consideration to integrating sensors into your robot's design; as competitions continue to increase in complexity, sensors may very well become an absolute necessity. Next year, you may find that sensors are no longer optional but standard equipment. Experiment and learn how to use sensors now so your team is ahead of the game this year and next.

13

GETTING ORGANIZED FOR PROGRAMMING

Get things organized before your team starts to program. Programming often gets complicated, and we're not talking about just the complexity of the work. Testing your programs requires keeping detailed records, and you can implement steps now, before a single NXT-G programming block is placed, to help your team stay organized and reduce confusion during the programming process.

During the course of building, programming, and testing a robot, your team will most likely build up a large collection of programs, some of which are specific to a single Robot Game mission and others for running multiple missions. Over time, the team will modify and combine the programs, moving and deleting entire sections of programming.

If you don't have a good way to track your work, the programs quickly end up as a big mess. That mess isn't just confusing though: It poses a danger to the success of your missions if you lose track of important programs. This chapter focuses on some indispensable housekeeping chores to build your programming arsenal. Examples include naming programs, using version numbers, and backing up your data, just to name a few.

If you're just not certain where or how to start programming, we offer a structured way to begin. We also cover flowcharts in this chapter, because a team sometimes needs a method not only for organizing existing programs, but also for organizing its thoughts and observations about the various missions before programming actually begins.

File-Naming Conventions

When you begin to create programs, save them with descriptive filenames that tell you something about each particular program so they are easy to find when you're digging through your program folder. For example, rather than call a program *push7.rbt*, written for the second mission of Building Team A's robot, use a more descriptive name, such as *RobotAMission2Final.rbt*.

From the filename *RobotAMission2Final.rbt,* we can infer that the program is probably designed for Robot A's approach to Mission #2 (these would be your naming conventions, not FLL's). The word *Final* suggests that, working or not, this is the finished version of the program.

Implementing an easy-to-understand file-naming system will save lots of time as the competition season moves forward. Whatever you do, decide on a file-naming system, and make certain that all the team members understand how to use it. Consider assigning the role of Program and Data Manager to a team member who will make certain that the program collection is properly named, tracked, and backed up (as discussed in Chapter 6).

Try the following conventions to see if they make organizing your programs a little easier.

Label Your Robots

If you are testing multiple robots, label each one; for example, *BotA*, *BotB*, and *BotC*. Once you select a single robot, remove this label from filenames. Teams could even develop labels for using specific attachments such as *BotA-Att1* or *BotC-Att4*. The key is to create labels that every team member can easily remember and understand.

Assign Mission Numbers or Names

Number all of the FLL missions, or give them short names. Make certain all the participants are in agreement on how to refer to each mission to avoid confusion. Examples include *M1*, *Mission3*, *LiftBarrels*, and *DragCar*.

Use Version Numbers and Status Words

When you create a program, you can safely assume that you will make changes as you test the program, and the team should track these updates accordingly. One method is to append *v1.0* (*version 1.0*) to the end of the original filename. Then, when you make a change to the program, save it using a new filename with an updated version number, such as *v2.0* (*version 2.0*). Don't ever overwrite a previous version of a program; archiving earlier versions of a program allows you to return to a previous version of a program if you find that a later version contains too many errors to fix.

Another option is to add status words to the end of a program name. *Status words* are short descriptions of a program's success or failure and might include *Failed, Test, Final,* and *Old.* Combine status words with version numbers for more descriptive filenames.

Create Longer, Descriptive Names

When naming your files, combine all of the above conventions to create longer, more descriptive names, rather than abbreviated ones. You can always rename the programs as competition nears so the shorter names are visible on the NXT LCD screen.

Examples of good program names that use the above conventions include the following:

- *BotAM4v2.0.rbt*
- *RobotCM5Final.rbt*
- *RobBAtt3Mission3.rbt*
- *RobAM2and4Test.rbt*

Back Up Your Programs and Data

Back up your programs and data. If your computer's hard drive fails, having a backup system in place will allow you to recover all of the team's hard work, including the robot's programs. Your team probably stores permission forms, worksheets, interview transcripts, and more in digital form (text documents, PowerPoint slide shows, and so on). If you don't back up this information and it is lost, your team could be out of the competition if time is short and the team can't recreate the work quickly enough.

Fortunately, backing up your programs and data has never been easier, whether you back up to a CD or DVD, a USB flash drive, or an online service.

The following suggested procedure ensures that your data is well protected:

1. Create a folder on a computer called *FLL Backups* to store all the team's data.
2. After each team gathering at which any type of computer work is done, make a copy of all the team's files to date, place them in a subfolder of *FLL Backups,* and name that subfolder using the current date (for example, *March_16*).
3. Copy all the contents of the *FLL Backups* folder to a flash drive or a writable CD/DVD.

Ideally, your team should use only a single computer or laptop to hold all its files, but this is rarely possible. Team members often work at home or in smaller groups when time permits. Because of this, files tend to be scattered among multiple computers. A Program and Data Manager can gather all the team files by email, USB flash drive, or various other ways and collect them on one machine.

Whatever method your team decides to use to protect its data, be sure to use it; a good backup plan is worthless if you don't implement it.

NOTE *Many teams purchase and create a team website from a web-hosting service, which often provides enough space to use as file storage for the team. Ask your hosting service about setting up an FTP service (and how to use it) and the software required to allow team members to store their individual work files. Storing your files with a web-hosting service is another way to back them up and keep them in one place that all the members can access. Another option is to purchase online backup storage space from services such as S3 by Amazon (http://aws.amazon.com/s3). Services such as this allow you to upload your data to storage servers and charge you by the megabyte or gigabyte of stored data, but it's typically affordable and easy to use.*

Robot Mission

After examining the competition mat and models, the team members typically understand what their robot must do; the challenge is programming the robot to accomplish its missions. The best way to begin is with a plan. Our recommended approach is to use flowcharts to structure the programs before actually beginning to write any programs.

Flowcharts are a graphical way of organizing the structure of a robot's program. You can find much more information on websites covering flowcharts using a simple Google search with keywords of *flowchart* and *tutorial*. A good website to start from is *http://www.hci.com.au/hcisite2/toolkit/flowchar.htm*, which contains one method for creating flowcharts, but this chapter demonstrates our flowchart method by creating a simple mission for the robot to perform. We show how to convert the mission description to a flowchart and then convert the flowchart to an actual NXT-G program.

Figure 13-1 shows the simple TriBot robot equipped with one Ultrasonic Sensor and one Touch Sensor.

For this mission, place the TriBot on the floor between two small boxes approximately 3 feet (1 meter) apart, with the TriBot facing one of the boxes. When the program starts, the robot should roll forward and stop when it is within 3 inches (7.62 cm) of the box. Again rolling forward, when the Touch Sensor is triggered, the robot should stop, rotate 180 degrees, roll forward, and stop within 3 inches (7.62 cm) of the second box. The robot's final movement will be to roll forward and stop when the second box triggers the Touch Sensor. Then the robot calculates the approximate distance (using degrees rotated) between the two boxes and displays the figure on the LCD screen.

Figure 13-1: A TriBot with two sensors, ready for action

Identify Individual Robot Actions

The next step is to break the mission down into individual actions by identifying every action of the robot and listing each action, as the following shows:

1. Roll forward.
2. Detect a box.
3. Stop forward movement.
4. Monitor Touch Sensor.
5. Rotate 180 degrees.
6. Calculate movement distance.
7. Display distance.

The robot will perform some of these actions more than once (such as detecting boxes 1 and 2) and will perform others only once (such as displaying the distance).

Having identified the individual actions, we describe each action, including any sensors, motors, or other special requirements needed to perform the action. (This often requires some brainstorming and discussion by the team; the more team members that participate, the less likely that something will be overlooked.)

The following is our sample mission actions with descriptions:

1. **Roll forward**   We might want to test to determine the optimal speed for the motors to rotate. Although the speed might not matter, it's a good idea to check to be certain.

2. **Detect a box**   Use the Ultrasonic Sensor to detect the first box. Mount the sensor low so it doesn't miss the box.

3. **Stop forward movement**   Use the Brake setting instead of Coast for more accuracy.

4. **Monitor Touch Sensor**   You may need to add a bumper to the Touch Sensor to ensure that it finds the second box (find information on bumpers in Chapter 10). Have the robot roll slowly toward the box.

5. **Rotate 180 degrees**   We need to calculate the correct number of rotations so the robot turns exactly 180 degrees.

6. **Calculate movement distance**   Monitor the number of rotations of one of the motors from the time it completes its rotation of 180 degrees until the second box triggers the Touch Sensor. For more accuracy, measure the distance from the first box to the back of the wheels, and add that value to the degrees/rotations it will take the robot to move to the second box.

7. **Display distance**   Measure the wheel's circumference to help calculate the actual distance traveled.

Don't skimp on the descriptions. The more information the team provides, the easier it will be to create the flowchart and actual NXT-G program. You will see why in the next section.

Convert Individual Actions to Icons

Next, convert each action into a small icon. Be as creative as you like or keep it simple, but have fun. Figure 13-2 shows the icons we created for our actions.

As you can see, the icons are nothing fancy; they simply need to represent the individual actions we identified from the robot mission description.

Arrange Icons to Mirror Mission Description

Print the icons on paper and cut them out, or use them in a graphics program. Following the mission description, place your icons, left to right, in the order described by the mission, as Figure 13-3 shows. (If you run out of space on a line, just wrap the icons to the next line.)

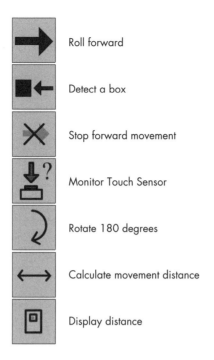

Roll forward

Detect a box

Stop forward movement

Monitor Touch Sensor

Rotate 180 degrees

Calculate movement distance

Display distance

Figure 13-2: Individual actions converted to icons

NOTE *You should add your own captions below the icons to explain what is happening in your mission descriptions, as Figure 13-3 shows.*

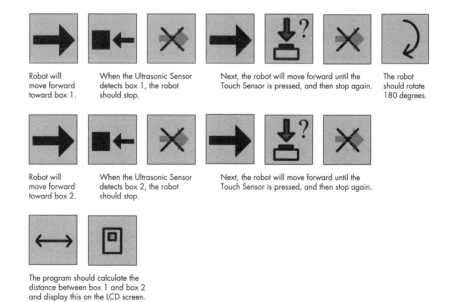

Robot will move forward toward box 1.

When the Ultrasonic Sensor detects box 1, the robot should stop.

Next, the robot will move forward until the Touch Sensor is pressed, and then stop again.

The robot should rotate 180 degrees.

Robot will move forward toward box 2.

When the Ultrasonic Sensor detects box 2, the robot should stop.

Next, the robot will move forward until the Touch Sensor is pressed, and then stop again.

The program should calculate the distance between box 1 and box 2 and display this on the LCD screen.

Figure 13-3: Icons match the order from the mission description.

Guess what? You just created a flowchart by converting the written mission description to graphic form. All that's left is to convert the flowchart to NXT-G and test it with a robot. The next section shows a partial conversion of this flowchart into an NXT-G program.

Convert Flowchart to NXT-G

Now we convert our icons into the actual NXT-G blocks required for the new program. At this point, we assume that anyone involved in programming the robot is familiar with using the NXT-G software. A good understanding of each NXT-G programming block is required; to convert an icon to its matching NXT-G block, you must know not only which NXT-G block to add to the program, but also how to configure it.

NOTE *Programmers should build the sample robots and complete the tutorials that come with the NXT-G software. Consult the Help documentation included with the NXT-G software for assistance on accessing the tutorials.*

To convert the icons, first open the NXT-G software. Begin by examining our icons and determining the proper NXT-G block (or blocks) to add to mirror the behavior described by each icon.

Figure 13-4: The robot will begin to move forward.

In the first part of our program, step 1, the robot moves forward until the Ultrasonic Sensor detects a box (step 2), at which point the robot stops (step 3). This sequence of events is programmed using a combination of Move and Wait blocks as Figures 13-4 and 13-5 show. First we added a Move block and configured it for Unlimited movement. This means the robot will roll forward until stopped (by an object, or in this case, the Ultrasonic Sensor detecting a box). In Figure 13-5, notice that we configured the block to wait until the Ultrasonic Sensor detects something (box 1) less than 3 inches in front of the robot.

Figure 13-5: The Ultrasonic Sensor will look for box 1.

Once the Ultrasonic Sensor is triggered, the Wait block stops "waiting" and lets the program continue. Figure 13-6 shows that after the Ultrasonic Sensor is triggered, the robot needs to stop (step 3).

To perform this action, we drop in another Move block configured to stop the motors, as Figure 13-6 shows.

Figure 13-6: The motors need to stop once the Ultrasonic Sensor detects box 1 (step 3).

Next, we want the robot to roll toward box 1 until it triggers the Touch Sensor (step 4). Our notes on this action tell us that we want the robot to roll slowly toward box 1 to keep from pushing the box. Accomplish this with another Move block (Figure 13-7) configured to move at a slower speed (lower power).

Figure 13-7: The robot will move slowly toward box 1.

Next, we want the robot to roll forward until the Touch Sensor is triggered (step 4). Use another Wait block that will "wait" until the Touch Sensor is pressed, as Figure 13-8 shows.

Figure 13-8: The robot moves slowly toward box 1 until the Touch Sensor is pressed (step 4).

Once the Touch Sensor is triggered, the robot should stop again (as it did earlier in step 3), as Figure 13-9 shows.

Figure 13-9: The robot stops when the Touch Sensor presses against box 1.

Referring to our numbered flowchart actions, in step 5, the robot should rotate 180 degrees and face box 2. To accomplish this, use a Move block as Figure 13-10 shows, and drag the Steering Control all the way to the right (motor B).

NOTE   *During testing, we would experiment with the degrees value to get the proper rotation, but for now we'll use a value of 500 degrees as a placeholder (to be tweaked later).*

Figure 13-10: The robot rotates 180 degrees to face box 2.

With the robot facing box 2, we realize the program is missing a step. During the programming phase, your team will frequently find that it has overlooked something. Remember that one of the mission objectives was to measure the distance between the two boxes. In our example, we forgot to include a way to record the distance traveled by the robot. We need a way for the robot to track the number of degrees that the robot's motors rotate between box 1 and box 2. Later we can convert degrees to inches (or centimeters) to determine the distance traveled.

NOTE   *If you print out your icons and place them on a tabletop, it's very easy to modify your flowchart by shifting icons left and right. Of course, you could also do the same on your computer.*

To add this step, add a new icon indicating that the robot should begin tracking the motors' degrees to the flowchart, as Figure 13-11 shows.

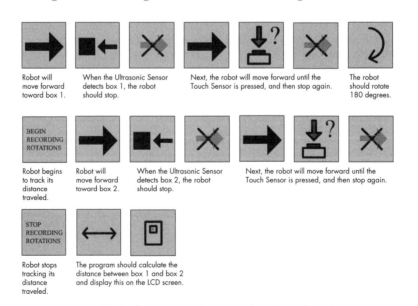

Figure 13-11: Modify the flowchart to show how the robot will track distance traveled.

How will we track this value? Assuming the programmers are familiar with all the NXT-G blocks and their functions, they will probably determine that the best block for this job is the Variable block. If we examine our flowchart carefully, we can see that we need to store two different values: value 1, for the degrees rotated from box 1 to the point the Ultrasonic Sensor detects box 2, and value 2, for the degrees rotated before the Touch Sensor detects box 2. In Figure 13-11, Variable blocks represent the icons for *Start Recording Rotations* and *Stop Recording Rotations*.

Having modified the flowchart in Figure 13-11, return to the program. To track the number of degrees a motor rotates, use a Rotation Sensor block. First place a Rotation Sensor block in our program (Figure 13-12), and configure it to record rotations for motor C (or just as easily set it for motor B). Also configure it to reset the counter so that the rotations will be recorded starting from 0.

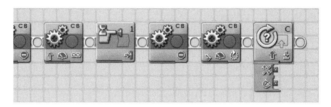

Figure 13-12: The Rotation Sensor block allows us to track the number of degrees of rotation for motor C.

Next, add a Move block, and configure it as Figure 13-13 shows. This Move block will start the robot moving toward box 2 while the Rotation Sensor inside motor C keeps track of how many degrees it rotates.

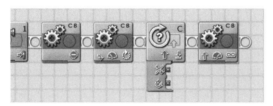

Figure 13-13: The robot begins to move again and tracks the distance traveled.

Your team's programmers can complete the remaining conversion of the flowchart icons to NXT-G blocks. We hope you're convinced that methods such as this one can be useful for breaking down complicated tasks (such as FLL missions). Once you understand those simple tasks, it's just a matter of finding the proper NXT-G blocks to make those tasks happen.

NOTE *Download the complete program for this chapter's mission, titled* ch13mission.rbt, *at* http://thenxtstep.com/book/downloads/.

Flowcharts Work

There is a lot more to flowchart design than we cover here. We hope you see that a little preplanning before actual hands-on programming occurs can save considerable time in the long run. By examining your robot's required actions on the competition mat, the team can develop a "map" to help guide the programming.

JUDGING TIP

If you're thinking, "I'm an expert with NXT-G. I can just jump from the mission description straight to NXT-G block programming," think again. Technical judges like to see that teams understand how they got from point A to point C; teams score points for showing that they understand the process. That's where flowcharting comes in; think of a flowchart as point B. If your team shows the technical judges how it analyzed the process of programming, the team is much more likely to get extra credit for documenting all that hard work. Also, a flowchart helps to dispel any concerns that a team received outside help from more technically competent persons.

NOTE   *We haven't even scratched the surface for creating icons for every NXT-G block (such as the Loop or Switch blocks). Your programmers might experiment with developing different types of icons for each sensor, motor, and NXT-G block, or they may simply create the icons as they need them.*

Summary

This chapter covers quite a bit. We hope that we show how preparation can save time and reduce stress. Programming is complicated enough without having to worry about hunting down lost programs, figuring out which program is the most current version, or even puzzling over how to start a complex mission.

At this point, your team should have its filename procedures in place and a data backup plan ready to go. You also have a strategy for dissecting missions and for developing a flowchart to use to map out the programming. Chapter 14 walks you through moving from flowchart to NXT-G program.

14

NXT-G PROGRAMMING CONCEPTS

Team members need a solid understanding of the basics of *NXT-G*, the drag-and-drop programming environment designed for the LEGO MINDSTORMS NXT robotics kit, to successfully compete in the Robot Game portion of the FLL competition.

For the Robot Game, the building of the robot frequently overshadows the programming. Nevertheless, it can be argued that programming is the more important skill; without it, your awesome robot will make a nice bookend.

This chapter first reviews some of the key NXT-G programming blocks (such as the Move and Loop blocks), discusses why they're important to your FLL robot, and offers some ways to use them in competition.

Next, we show how to use what we call the *secret power blocks*: Variable and File Access. You may not be as familiar with these blocks as you are with the Move, Loop, Switch, and Wait blocks, but you'll be surprised to discover how useful they can be.

Then we present and discuss some samples of actual NXT-G programs and demonstrate how they can benefit a robot, including how best to use and program the Light Sensor, Ultrasonic Sensor, Touch Sensor, and the built-in timers.

While we can't show you how to do everything with NXT-G (that would take the fun out of experimenting!), we hope this chapter provides you with a nice selection of ideas to consider and build on. So let's give that robot something to do and dive right in.

NOTE *Teaching how to program with NXT-G in detail is beyond the scope of this book. For more on programming NXT-G, consult one of the books on the topic, and review the tutorials that come with your NXT-G software. We suggest both* The LEGO MINDSTORMS NXT Idea Book *by the contributors to The NXT STEP Blog (No Starch Press, 2007) and* LEGO MINDSTORMS NXT-G Programming Guide *by James Floyd Kelly (Apress, 2007). Familiarize yourself with all the NXT-G programming blocks, their respective configuration panels, and the processes involved with uploading and deleting programs from the NXT Brick.*

Key NXT-G Blocks

It's probably a safe bet that every NXT-G block available is on someone's list of *must-have* blocks. Our own list includes the following: the Move, Loop, Switch, and Wait blocks. But this doesn't mean that your team shouldn't learn the details of each and every NXT-G block available; the more blocks the team completely understands, the more powerful your programs can become.

We highly recommend that every team member be familiar with every block used in the team robot's program(s), because technical judges often like to see cross training on a team. The programmers need to understand the how and why of the robot design, and every team member should be able to answer questions about the team Project.

TESTING A ROBOT'S PHYSICAL ABILITIES

NXT-G's true programming power comes from experimentation. When developing your robot and its programs, be patient and examine what the robot must physically accomplish; if you can "drive" your robot around the mat by hand, using its chassis and attachments to complete a mission, you can create a program for the robot to perform the same task autonomously. On the other hand, if you find that you cannot complete a mission by test-driving the robot by hand, then no amount of programming will solve the problem. At that point, take a step back, examine the problem again, and start fresh, knowing that you learned something new.

Move

A Move block (Figure 14-1) not only allows you to control the speed, degrees or rotations to turn, time to spin, direction, and braking of the NXT motors, but you can also select which motors (A, B, or C) the block will control. Without the Move block, your robot probably won't be competing in the Robot Game; it is probably the most used NXT-G block, and in fact, many teams have programmed robots using nothing but Move blocks!

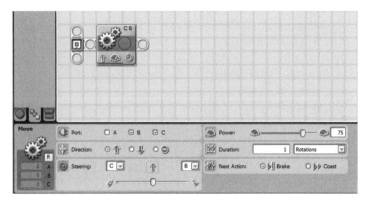

Figure 14-1: The Move block and its configuration panel

Some uses of the Move block include the following:

- Quickly stop your robot using the Brake option on the Move block's configuration panel. The Coast option makes it difficult to program precise movements; when power to the motor stops, the wheels still rotate a bit as the robot's momentum carries it farther along. Use the Coast option by allowing the robot to coast when returning to Base to save battery power.

- When your robot is in a tight space, use the Steering option on the Move block's configuration panel to allow the robot to spin in place with no forward or backward movement; simply drag the slider all the way to the left or right.

- During mission testing, lower the Move block's Power setting; this saves battery power and lets you carefully observe the robot's movements. But remember to readjust the power to reduce the mission's completion time. Careful testing will provide the right Power setting for each mission. You will see a change of a decrease or increase in speed when dragging the Power setting left or right, respectively.

- The Move block can use data wires to communicate with other blocks. When your robot rotates in one direction, consider sending that value to a Variable block, wire that Variable block to another Move block, and have the robot rotate the exact number of degrees/rotations in reverse. This can be useful for getting a robot to a very specific location and back again using recorded directional values. (See "Secret Power Blocks: Variable and File Access " on page 176 for more information on the Variable block.)

Loop

A Loop block (Figure 14-2) allows a program to repeat the execution of any blocks inside itself; this is called *looping*. Looping can last forever (or until the batteries give out), for a specified number of repeats, or for as long as a certain condition exists (for example, if the Light Sensor block inside a Loop block detects a black line on a white surface, the loop will keep repeating). When the loop stops executing, this is called "breaking the loop." If, for example, the Light Sensor detects the white surface instead of the line, the loop will stop repeating, or break.

Teams often ignore the Loop block because most Robot Game missions do not have repeatable activities. If, for example, a mission involves a robot delivering one object at a time from point A to point B, the use of a Loop block would be quite obvious. But even though missions do not typically include these types of actions, this doesn't mean the Loop block isn't useful.

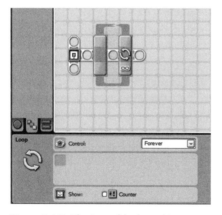

Figure 14-2: The Loop block and its configuration panel

You might find the Loop block useful in FLL in some of the following ways:

- When testing a mission in which the robot returns to Base, place all the NXT-G blocks for the program inside a single Loop block configured to run Forever. Add a Wait block as the final block (inside the Loop), and configure it to wait for 5 or 10 seconds. As a result, you won't need to select and execute the program again to run the mission; just spin the robot around at Base, and in 5 to 10 seconds, it will keep running the mission until you cancel the program.

- A Loop block is useful for making certain your robot has accomplished a specific movement. For example, if the robot needs to pull a lever, consider putting the Move block required inside a Loop block configured to repeat three times. When the robot reaches the lever, it will attempt the pull three times instead of once, improving the chances that it successfully pulls the lever. (This is also useful for pushing an item against an object or grabbing/collecting a group of objects.)

Switch

The Switch block (Figure 14-3) is one of the most useful blocks for giving your robot the ability to make a selection from multiple options. When paired with sensor input, the Switch block provides real decision-making capabilities. Even better, the Switch block can make decisions based on logical input (such as true/false) and number values (motor rotation or light level, for example). But these are common uses of the Switch block; we discuss some more unique uses below.

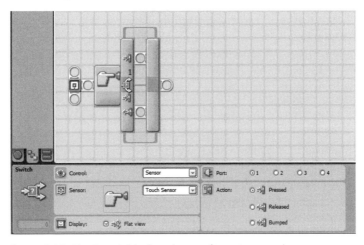

Figure 14-3: The Switch block and its configuration panel

Create a Light Sensor Toggle

Did you know that you can configure a Light Sensor as a toggle for running one of two possible subprograms (Prog1 and Prog2)? Figure 14-4 shows how simple it is to implement such a Light Sensor toggle using a Switch block. In the upper path of the Switch block, we dropped in a Move block (Prog1). In the lower path, we placed an Ultrasonic Sensor block (Prog2). Depending on whether the Light Sensor is covered or uncovered will determine which subprogram the robot executes.

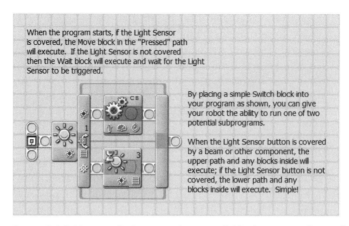

When the program starts, if the Light Sensor is covered, the Move block in the "Pressed" path will execute. If the Light Sensor is not covered then the Wait block will execute and wait for the Light Sensor to be triggered.

By placing a simple Switch block into your program as shown, you can give your robot the ability to run one of two potential subprograms.

When the Light Sensor button is covered by a beam or other component, the upper path and any blocks inside will execute; if the Light Sensor button is not covered, the lower path and any blocks inside will execute. Simple!

Figure 14-4: Use a Light Sensor with a Switch block as a simple toggle.

A Light Sensor is normally added to a robot in an uncovered state; that is, the sensor's light-sensitive element is unobstructed and allowed to detect the ambient light level of its surroundings. The program in Figure 14-4 is configured to execute any blocks in the Switch block's upper path if the Light Sensor detects a light value greater than 10 (in a lit room). If the Light Sensor detects a light value less than 10 (covered or in a dark room), any blocks in the lower path will execute.

With the Light Sensor in a lit room and an uncovered state and the program is executed, the Switch's upper path in Figure 14-4 will always run. If we configure the Light Sensor as starting completely covered by using some TECHNIC components to obstruct the light-sensitive element, we can force the program to always execute the lower path. The secret to using the Light Sensor toggle is to have the robot execute Prog1 with the Light Sensor starting in an obstructed state. Program the robot to return to Base when it completes Prog1. When the robot returns to Base, remove the TECHNIC pieces covering the Light Sensor, and ready the robot to complete the next mission using Prog2.

NOTE   *Download this sample NXT-G program* toggle.rbt *at* http://thenxtstep.com/book/downloads/.

Build a Menu System

The Switch block and Touch Sensor toggle device work very nicely together when your robot needs to run only two programs. But what if you have three or more programs? You can build a menu system (see Figure 14-5)!

Many teams build a menu system that allows them to select from a list of mission subprograms (such as Prog1 and Prog2 above). If you have a relatively small list (for example, four missions), the menu system could simply use the Left and Right buttons to select the mission number via the Brick's LCD.

What if you have more than four missions? The menu system contains all of the blocks used to create the menu system and all the NXT-G blocks for every subprogram. This means that for more than three missions, your program can become very large, so keep an eye on memory usage.

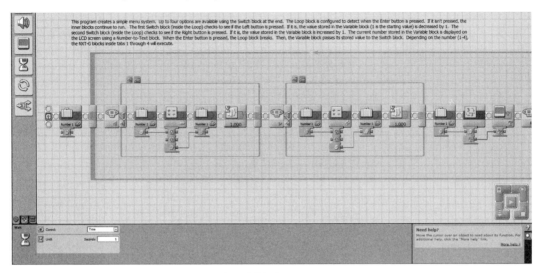

Figure 14-5: Create a menu system using a Switch block.

Download, review, and modify the NXT-G blocks used to create the *menu.rbt* program that Figure 14-5 shows (along with comments) from *http:// thenxtstep.com/book/downloads/*. Keep in mind that this sample does not handle errors (if there are four programs and the user chooses five, the program won't know what to do), and you cannot change a selection. We leave it to you to experiment and provide those NXT-G blocks.

Wait

The Wait block (Figure 14-6) is fairly self-explanatory. If you drop it into your programs, the program will pause until either a time limit expires or a sensor is triggered.

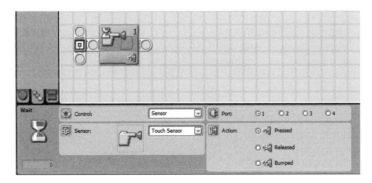

Figure 14-6: The Wait block and its configuration panel

Consider adding the following uses of the Wait block to your programs:

- If your team has an extra Touch Sensor, consider connecting it to your robot as a Start button, and then drop in a Wait block configured to trigger when the Touch Sensor button is Bumped. Why would you want to use a Touch Sensor as a Start button? Frequently, when team members press the orange **Enter** button, they must quickly pull back their hands. This can cause problems when the robot immediately begins to move—sometimes the robot is knocked off course when an attachment or component impacts a hand. Placing the Touch Sensor to the side or rear of the robot and using the Wait block reduces that risk.

- During testing, the Wait block can be useful for taking measurements and evaluating the robot's performance on the mat. At various points in your programs, place Wait blocks that pause the program until a Touch Sensor or NXT button is Pressed. By using this method, the robot will wait as the team monitors its movements. When it's time for the next step, trigger the Wait block, and the robot will continue until the program reaches the next Wait block. Continue this process as needed.

Secret Power Blocks: Variable and File Access

These blocks aren't really secrets, but the Variable and File Access blocks are rarely seen in NXT-G competition. The goal of this section is to demonstrate how these blocks can provide more flexibility in your robot's programs.

Variable Blocks

The Variable block (Figure 14-7) can hold one of three types of data: logic, number, or text. We focus on using the block to hold numbers, but the same methods apply to logic and text data types.

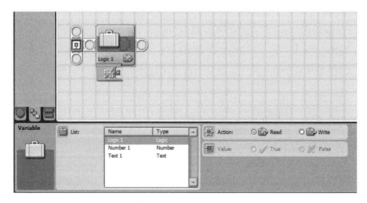

Figure 14-7: The Variable block and its configuration panel

When using a Variable block, create your own variable names. NXT-G comes with three built-in variables, *Number1*, *Text1*, and *Logic1*, but your own variable names will probably be more meaningful to you. For example, the names *MotorBRotation* and *DegreesRight* immediately tell you something about their functions.

Create a New Variable

To create a new variable, use the following steps:

1. Select **Edit ▶ Define Variables** (Figure 14-8).
2. In the Edit Variables window, click the **Create** button (Figure 14-9).
3. Enter a name for the variable in the Name text box, and select a variable type from the drop-down window (Figure 14-10).
4. Click the **Close** button when finished (or click the **Create** button again to create a new variable).

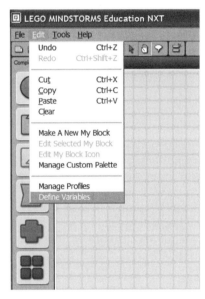

Figure 14-8: Select Define Variables to create a new variable.

Figure 14-9: Click the Create button to add your new variable.

Figure 14-10: Your new variable is now available for use in your programs.

Use Your Variable Block

Consider the mission that Figure 14-11 shows. Your robot needs to deliver eight small items to an area on the mat designated by a small gray square, with an unknown distance from Base. Only items that rest completely inside the square will receive points, and the robot can deliver only one item at a time, after which it must return to Base for the next item.

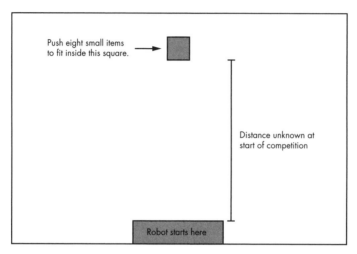

Figure 14-11: The robot must deliver items to the small square.

To accomplish this mission, many teams use a Light Sensor to detect the front edge of the square. To keep from overshooting the small square, they reduce the robot's speed so the Light Sensor can detect the edge and quickly stop the robot. The robot slowly moves forward a small distance to deliver its item.

This solution works, but it is time consuming (if the robot's speed is slow), and it requires using the Light Sensor to look for the small square every time it approaches. Also, the team may run out of time before delivering all eight items. The good news is that there's a faster method.

First, have the robot use the Light Sensor to detect the front edge of the square, but only once. While the robot is moving toward the square, track the number of degrees one of the robot's motors rotates, and write this to a Variable block called *DisToSquare*. We showed how to use the Rotation Sensor block in Chapter 13 to monitor the number of rotations/degrees that a motor spins, so integrate this into the program that Figure 14-14 shows. After the robot's first run, the *DisToSquare* variable will contain the exact number of degrees for the motors to rotate.

Next, take the value stored in the variable and send it to the Move block that controls the robot's forward movement (also use this same value to have the robot move in reverse to return to Base). This is done inside the Loop block that Figure 14-12 shows.

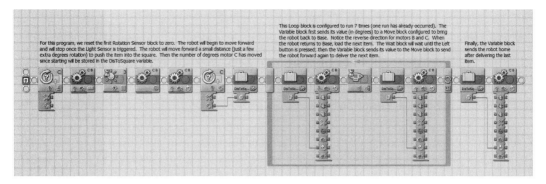

For this program, we reset the first Rotation Sensor block to zero. The robot will begin to move forward and will stop once the Light Sensor is triggered. The robot will move forward a small distance (just a few extra degrees rotation) to push the item into the square. Then the number of degrees motor C has moved since starting will be stored in the DisToSquare variable.

This Loop block is configured to run 7 times (one run has already occurred). The Variable block first sends its value (in degrees) to a Move block configured to bring the robot back to Base. Notice the reverse direction for motors B and C. When the robot returns to Base, load the next item. The Wait block will wait until the Left button is pressed; then the Variable block sends its value to the Move block to send the robot forward again to deliver the next item.

Finally, the Variable block sends the robot home after delivering the last item.

Figure 14-12: Using a Variable block to store a distance traveled by the robot

The Variable block gives the robot the ability to store a value that is determined during an actual mission and not before (in this example, the distance to the small square, which is unknown before the mission starts). Another example of a variable value that might change is when the Light Sensor needs to record a value for the color of an item (white, gray, black, and so on) that changes or is not known before the competition.

NOTE *Download this sample NXT-G program* square.rbt *at* http://thenxtstep.com/book/downloads/.

A More Complex Mission

Before moving on, let's consider a variation of this mission. Let's say the mat contains four squares (as Figure 14-13 shows). The robot needs to deliver 16 items with 4 items per square and must load each square with an equal number of items (that is, squares 1, 2, 3, and 4 must all contain one item before adding a second, and then the squares must all contain two items before adding a third item, and so on).

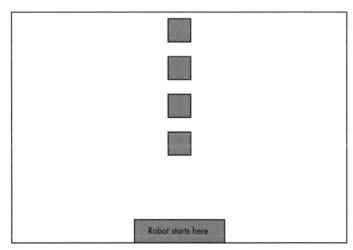

Robot starts here

Figure 14-13: The robot must deliver items to all four squares.

We could use four Variable blocks, with each variable holding the distance from one square to the next. However, if you look at how often the single Variable block is used in the program that Figure 14-12 shows, you can imagine the large size of the program if we used four variable blocks! Luckily, there is another method available.

Modified Mission Solution

In this modified mission, the robot will move straight toward the first square, drop its item, and then move on to the second square, then the third, and finally the fourth. The robot then returns to Base for the next four items. It will repeat the above movement three more times, placing a total of 16 items on the mat.

Ideally, we want the robot to remember how far it moves forward to squares 1, 2, 3, and 4 and then repeat these distances for three additional runs. To do this, we use the File Access block.

The File Access Block

The File Access block is similar to the Variable block, except that, unlike the Variable block, it can hold multiple values by storing the information in a file on the Brick.

The File Access block can perform four actions: Write, Read, Close, and Delete (as the pop-up menu in Figure 14-14 shows). The Write action allows the robot to put a piece of data (text or number) into the file. Read allows it to read an existing piece of data from a file. Before your robot can read data from a file, the file must be closed using the Close action. Finally, your robot can remove a file stored on the Brick by using the Delete action.

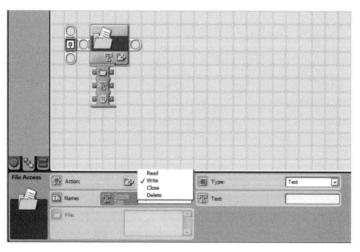

Figure 14-14: The File Access block has four options.

If you're wondering how a File Access block holds the robot's data, it's fairly simple. Data is stored sequentially in a file, from first to last. So, if the robot were to write the number of degrees a motor rotates as it travels from square to square to the File Access block, the data in the file might look like this:

$$720, 680, 820, 940$$

NOTE *The commas don't exist in the file but are used here to differentiate between the four values. The value of 720 was the first number of degrees that was read by the Rotation Sensor and written to the File Access block, followed by 680, 820, and finally 940.*

Running Our Modified Mission

As with the first mission, use the Light Sensor to detect all four squares. Reset the Rotation Sensor block between squares, and as each square is discovered, write the Rotation Sensor value to a file using the File Access block. Finally, when the robot reaches the last square, sum up the previous Rotation Sensor values (to know how many degrees of motor rotation it will take to get the robot back to Base), and store that in a Variable block.

NOTE *If your robot rolls over the squares, it will probably roll over items left in its path to and from Base and will be knocked off course. You must figure out how to deliver the items so the robot doesn't roll over items left in squares. (Hint: the robot itself doesn't have to roll over the squares, just the Light Sensor.)*

Figure 14-15 shows part of the program that accomplishes this task. Download and examine the entire program to see how it works, but for now we'll point out just the part where the File Access block writes and reads from a file.

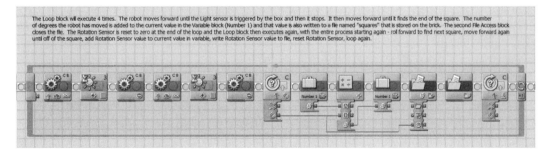

Figure 14-15: The File Access block writes Rotation Sensor values to a file.

Each time the Rotation Sensor writes its value to the File Access block, the file is closed (using the second File Access block inside the loop as Figure 14-15 shows). Also add the value from the Rotation Sensor to a Variable block (Number 1) that just totals the complete distance traveled. Use the Variable block to return the robot to Base. There's a lot happening in Figure 14-15, so the following is a breakdown and simple description of the process, from left to right:

1. The Rotation Sensor block resets its internal value to zero.

2. The Loop executes 4 times—the 13 blocks inside measure the distances between the four squares and write them to a file stored on the Brick.

3. The first Move block starts the robot rolling, while the Light Sensor block looks for the start of a square. When it finds one, another Move block stops the robot.

4. Next, one more Move block slowly moves the robot forward to find the end of a square, where the robot stops again.

5. The value obtained by the Rotation Sensor is read (using a data wire) by a Math block. The Math block also reads any value currently stored in the Variable block and adds it to the current Rotation Sensor value.

6. The Rotation Sensor's value is also written to a file using the File Access block.

7. The File Access block closes the file, and the Rotation Sensor resets to zero.

8. The loop executes again (unless it was the fourth loop, when it breaks the loop and continues with the rest of the program).

Figure 14-16 shows the next portion of the program that reads the Variable block and reverses the robot's direction to return it to Base. Remember, the Variable block should now contain a number that equals the total number of degrees the robot moved forward from its starting position to the end of the last square.

Now take the Light Sensor out of the process. Four values stored in a file on our Brick contain the number of degrees the motors should rotate to return to the correct points on the mat. Using a combination of Loop blocks (as Figure 14-16 shows), the robot will return three more times to drop off items at the four squares (it already dropped one item off in all four squares during the first run when it gathered the distance data, right?). This time, however, the robot can move faster since it is not relying on the Light Sensor.

The following is a brief description of the actions accomplished by the program elements in Figure 14-16.

1. The Variable block provides the total number of degrees the motors rotated from the starting position to the end of the last square. The Move block simply reverses the direction of motors B and C, and the robot ends up back at the starting point.

2. The outer Loop block and any blocks inside will execute a total of three times. Notice that there's an inner Loop block, too—we cover that in a moment.

3. Now that the robot is back at the starting position, the Wait block pauses the program until the Left button on the Brick is pressed (change this to any input you like for restarting the program).

4. The inner Loop block shown in Figure 14-16 will execute four times. Each loop corresponds to the robot moving forward toward a square.

5. As the robot moves forward, the File Access block begins to read from the start of the file. The first value stored in the file corresponds to the distance the robot must travel to reach square 1. The Move block reads that value, moves the robot so it can deposit an item on the square, and then the process loops again (a maximum of four times—step 5 repeats by reading the second, third, and last values stored in the file to get the robot to squares 2, 3, and 4).

6. After the inner Loop block executes four times, the robot is at the last square. The File Access block closes the file (so future reads will begin at the beginning of the file), and the Move block once again reads the Variable block's value so the robot can return to the starting position.

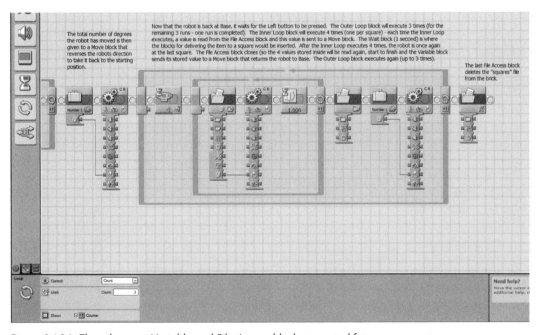

Figure 14-16: The robot uses Variable and File Access blocks to control future movements.

The secret power blocks may or may not come in handy for your robot. But you should now understand how to use them to simplify repetitive actions and hold variable data that you cannot determine prior to a competition. Add them to your growing "toolbox" of programming skills.

NOTE *Download this sample NXT-G program* squares.rbt *at* http://thenxtstep.com/book/downloads/.

Sensor Programming Techniques

This chapter concludes with some sample programs that show a few NXT sensors in action—Light, Touch, and Ultrasonic—as well as how to use the three built-in timers.

Light Sensor Range Reading

Let's look at how to program the Light Sensor to detect a specific value that has an upper and a lower range. If the Light Sensor can only detect whether the current reading is above or below a value, how would you, for example, program your robot to react if the Light Sensor value is less than 80 but greater than 60? See Figure 14-17 for the solution.

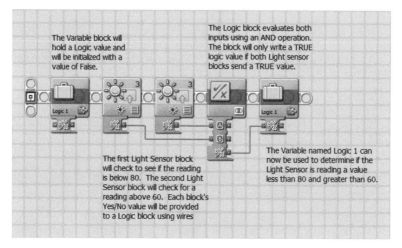

The Variable block will hold a Logic value and will be initialized with a value of False.

The Logic block evaluates both inputs using an AND operation. The block will only write a TRUE logic value if both Light sensor blocks send a TRUE value.

The first Light Sensor block will check to see if the reading is below 80. The second Light Sensor block will check for a reading above 60. Each block's Yes/No value will be provided to a Logic block using wires

The Variable named Logic 1 can now be used to determine if the Light Sensor is reading a value less than 80 and greater than 60.

Figure 14-17: Check to see if the Light Sensor reading falls between an upper and lower value.

In this example, we feed the *Yes/No* values from the two Light Sensor blocks to a Logic block. The Logic block uses an AND operation to determine if both Light Sensor blocks are providing true values. If so, the Logic block writes a value of *True* to the Variable block. Otherwise, the Variable block continues to hold a value of *False*.

You could use this small collection of blocks to give your robot the ability to follow a line that is a specific color. To do so, simply modify the sample Light Sensor threshold values of 80 and 60 (in Figure 14-17) by testing the line color you want the robot to follow and changing the upper and lower thresholds, respectively.

Using the Touch Sensor to Combine Two Programs

The next program (in Figure 14-18) demonstrates how to use the Touch Sensor to choose which of two programs to run.

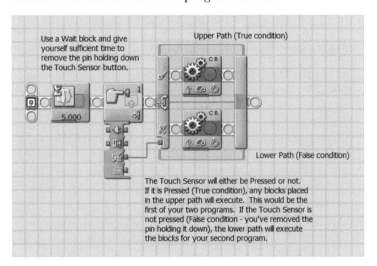

Figure 14-18: Use the Touch Sensor to select between two possible options.

In Figure 14-18, start by adding a Wait block with enough time for someone to remove the pin holding down the Touch Sensor; in this case, five seconds. Add a Touch Sensor block that checks whether the Touch Sensor is Pressed or Released. If it is Pressed, a logic value of *True* is sent to the Switch block, which is configured to use a logic value to determine whether to execute blocks in the upper or lower paths. We also placed one Move block in the upper path and another in the lower path.

When the program is executed and the pin isn't removed within five seconds, the Touch Sensor will be in the Pressed condition, and the Switch block will execute any blocks in the upper path. If the pin is removed, the Touch Sensor button will be in the Released condition, which triggers the Switch block to execute blocks in the lower path.

Since our goal is to combine two programs stored on the NXT Brick into one program, first copy the blocks from the first program into the upper path. Then copy the blocks from the second program into the lower path.

One method for using this program is to place the robot so it attempts the first mission using the blocks in the upper path of the Switch block. The upper path should also contain blocks to return the robot to Base at the end of the mission. When the robot reaches Base, you have five seconds to pick up the robot, position it for the second mission, and remove the pin holding down the Touch Sensor. If you don't remove the pin, it will attempt the first mission again—useful if the first mission fails and you want to try it again.

NOTE   *Be sure to add blocks to return your robot to Base if it needs to run more missions.*

Creating an Ultrasonic Sensor Toggle

Using the Ultrasonic Sensor in competition can be tricky. If another team is using the same sensor, the robots may be confused by all the radio waves bouncing around. Therefore, it's always good to mount the Ultrasonic Sensor as low as possible so its radio waves bounce off the Robot Table walls.

If the team decides not to use the Ultrasonic Sensor in competition because of potential interference from another robot sensor, there is another use besides object detection and avoidance. Figures 14-19 and 14-20 demonstrate an Ultrasonic Sensor toggle. Once again, we use a sensor as a means of program selection.

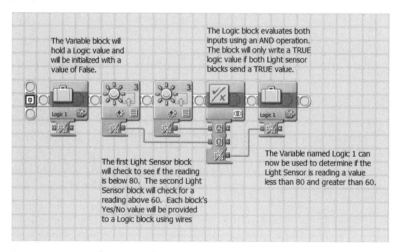

Figure 14-19: Configure an Ultrasonic Sensor as a toggle.

In this example, we built a toggle similar to the one that uses a Touch Sensor. Figure 14-19 shows the robot with the Ultrasonic Sensor blocked on the left and unblocked on the right. When the program starts, the robot waits for the Touch Sensor to be pressed, and then the Ultrasonic Sensor takes a reading. If its signal is blocked, the blocks in the upper/True path of The Switch Block Execute. If The Signal Is Not Blocked, Any Blocks In The Lower/false path of the Switch block execute.

NOTE   *Download the sample NXT-G program ultratoggle.rbt at* http://thenxtstep.com/book/downloads/.

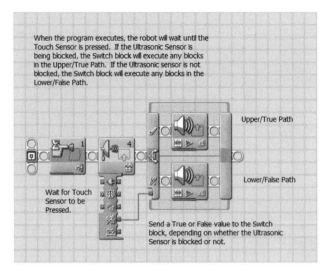

When the program executes, the robot will wait until the Touch Sensor is pressed. If the Ultrasonic Sensor is being blocked, the Switch block will execute any blocks in the Upper/True Path. If the Ultrasonic sensor is not blocked, the Switch block will execute any blocks in the Lower/False Path.

Upper/True Path

Lower/False Path

Wait for Touch Sensor to be Pressed.

Send a True or False value to the Switch block, depending on whether the Ultrasonic Sensor is blocked or not.

Figure 14-20: The program will check to see if the Ultrasonic Sensor is blocked or unblocked.

Using Timers for Testing Missions

Chapter 12 mentions that during testing, it can be helpful to use the NXT Brick's three internal timers to track how long the robot takes to perform a mission, from leaving Base to returning. This information could help you decide whether to increase motor speeds, combine mission programs, or possibly rank the order of attempted missions based on the ratio of possible points to the time required to complete the mission.

Figure 14-21 shows one set of NXT-G blocks that you could add to your programs. When you finish testing, simply remove the blocks to reduce the program size.

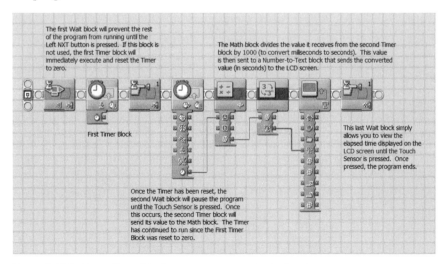

The first Wait block will prevent the rest of the program from running until the Left NXT button is pressed. If this block is not used, the first Timer block will immediately execute and reset the Timer to zero.

The Math block divides the value it receives from the second Timer block by 1000 (to convert milliseconds to seconds). This value is then sent to a Number-to-Text block that sends the converted value (in seconds) to the LCD screen.

First Timer Block

This last Wait block simply allows you to view the elapsed time displayed on the LCD screen until the Touch Sensor is pressed. Once pressed, the program ends.

Once the Timer has been reset, the second Wait block will pause the program until the Touch Sensor is pressed. Once this occurs, the second Timer block will send its value to the Math block. The Timer has continued to run since the First Timer Block was reset to zero.

Figure 14-21: Use a Timer block to measure a mission's duration.

In Figure 14-21, you would place any blocks your robot requires to complete its mission (including blocks to return it to Base) between the two Timer blocks. When the robot returns to Base, press the Touch Sensor button, and the LCD screen will display the time the mission took to complete.

You can choose to replace the Wait blocks (configured to use the NXT Brick's left button and the Touch Sensor) with any other method to trigger the reset of the timer, the end of the timer cycle, and the end of the program.

NOTE *Download the sample NXT-G program* timers.rbt *at* http://thenxtstep.com/book/downloads/.

Summary

We hope that this chapter gives you a sense not only of the programming power available to you via NXT-G but also of some potential design and programming techniques that you can immediately integrate into your Robot Game. The FLL competition changes every year, so rest assured that as the Robot Table missions change, so will your robot and its programs.

I have not failed. I've just found 10,000 ways that won't work.

—Thomas Alva Edison

15

THE PROJECT

This book spends a lot of time on the Robot Game,
but we haven't forgotten the Project! We'll discuss the
Project in this chapter.

For the Champion's Award, the Project counts as much as the Robot
Game. Since it varies widely from year to year, we won't discuss it at the level
of detail we discussed the Robot Game. However, we can discuss some of its
main components and make some general suggestions for ways to tackle it.

The Project usually consists of three main parts: research, community
outreach, and a presentation to judges. During the season, the teams research
and solve a real-world problem and create a presentation based on the Project.
A team is also encouraged to share its research and solutions with the community in some way. At a tournament, each team presents before a panel of
judges and answers questions.

Choose a Topic

Teams can choose specific topics based on a general theme. For example,
for the Nano Quest season, the Project was based on nanotechnology. As
part of the Project, teams had to choose and research an application of
nanotechnology, including scientific developments and the challenges

scientists face. Because the field of nanotechnology is broad, teams could choose many different topics. For example, one team might have chosen to research an application of nanotechnology in medicine, while another could have researched nanorobotics.

Before deciding on a research topic, consider having each member research a variety of topics related to the Project theme so you have a better idea of which one to pick. You may want to have only members of the Research Team (discussed in Chapter 6) do this preliminary research, depending on how much work the Building Team has and how many topics you want to research.

NOTE   *FLL themes are always fairly large in scope. The theme of alternative power, for example, can cover a variety of subjects such as solar, wind, ethanol, and so on. When selecting your topic, try not to choose one that's too generic (such as home solar panels) because you will likely find that other teams selected the same topic. Think differently! Have the team brainstorm about some truly wild ideas. For example, instead of researching home solar panels, how about researching a solar panel jacket that can charge your cell phone and other portable devices? Stand out as a team by picking a topic that is unique and interesting but also extremely fun!*

Once your team members finish researching the topics, have each give a short presentation of the findings to the entire team. The team should discuss each topic, perhaps suggest a few more, and eventually vote for the final one (a secret ballot is a good idea here). Feel free to change the topic if the team finds it isn't as good as originally thought.

However you choose a topic, include all the members in the decision-making process to keep the group happy.

Research the Topic

Once you decide on a topic, it's time to begin the research. Research can take many different forms, from researching on the Internet to doing your own experiments. A great place to start your research is on the FLL challenge website, where you should find numerous Project resources. These resources commonly include links to helpful websites and a list of helpful books and videos.

During the research phase, include as many team members as possible. Remember that the judges will ask all the participants about the research topic during the Project judging session, so be sure that they all have a good grasp of it.

Assign Tasks

When researching, consider splitting the tasks among several members. For example, if the team needs to research the advantages and disadvantages of solar power for a building, a couple of members might find the average cost of power purchased from an electric utility company, while another could look up the average cost of solar power, and other members could be assigned to look up the advantages and disadvantages of each source. This way, each member has a specific, manageable task.

The team Captain or the Research Team Leader can determine assignments, or the team members can decide. Either way, let the members discuss which assignments they'd like to tackle, and make your final assignments with their desires in mind.

Record Your Findings

When doing research, be sure to document your findings well. For example, you might have a fantastic interview with an expert, but then forget what was said when you prepare your presentation! Be sure to write everything down or record your interviews.

Similarly, when you find useful web articles, bookmark them or copy the article in a Word document so you have easy access to it later. You may find it helpful to have a single document in which you keep all your notes and findings, with references to sources. This not only helps you quickly find a previously researched item, but it can also help you organize it into a presentation.

Document Your Sources

It's important to keep track of your sources so you can give credit in your presentation. This is one of the items in the rubrics, so the judges will probably look to see how many sources you cite in your presentation.

In addition, citing sources enhances your presentation. For example, simply stating, "Solar power costs 58 percent more than power from the electric utility," doesn't sound as professional or convincing as, "The National Power Research Institution conducted a study in 2006 on the costs of different power options. This study indicated that solar power costs 58 percent more than power from the electric utility." You don't need a formal bibliography, just something that shows the source of the information included in your presentation.

Research Using the Internet

The Internet makes it easy to find information about a subject. But how do you begin to research? The following are some helpful resources that many teams find useful for research.

Search Engines

One of the most obvious information resources is an Internet search engine such as Yahoo! or Google. Simply type the subject of your research, and thousands of links usually come up in response (though many will be irrelevant). Many times, you can find a website out there that has just what you need!

Wikipedia

Online encyclopedias such as Wikipedia (*http://www.wikipedia.org/*) can also provide useful information. Wikipedia has a huge collection of articles on many different subjects; simply do a search for your subject. Typically, Wikipedia is best for learning about general subjects (such as nanotechnology), but it can also be very helpful with finding more specific information.

Ask-a-Scientist

Some websites have a feature called *ask-a-scientist*, where you can submit questions to real scientists! If you can't find what you're looking for by searching the Internet, you could try using one of these websites to get more specific information. For example, suppose you were wondering about the ideal shape of a solar panel. If you couldn't find this from other sources, you could submit a question on an ask-a-scientist website. One such site is *http://www.madsci.org/*, but you can find many ask-a-scientist websites simply by searching for *ask a scientist*.

Personal Interviews

Sometimes it helps to interview scientists or other people about the Project topic. In addition to helping you get more specific information, interviewing can improve the team's score, since one of the rubric items includes talking to science professionals.

A great way to find experts is to call or write to local universities. Call a specific department (such as engineering or physics), and briefly explain about FLL, the team, and the Project. Secretaries in these departments will often help you find the correct person to speak with and help set up interviews. And don't worry about distance; you can always interview someone over the phone.

You may also be able to interview people at businesses that sell related products. Most businesses are happy to help students and may be able to provide photos, technical information, and more. When your team calls businesses, always be polite, don't ask for too much help, and don't overstay your welcome. If one business won't help you, try another one.

Libraries

Libraries are, of course, great places to begin research. Find lots of up-to-date information by looking at technical journals and, in most libraries, a reference librarian who is trained in doing research. You might also be able to get a subscription to online books. Most libraries participate in a system that loans books, videos, and other resources between one another. If you can't find what you need at your local branch, ask a librarian to show you how to request resources from other libraries.

Field Trips

Field trips are probably one of the more exciting ways to do research. Besides being fun and great for team building, they can give the team a lot of help with background material about the topic. For example, if the Project is about space shuttles, your team could take a field trip to the Smithsonian National Air and Space Museum to learn some background information about shuttles and space exploration in general. Be sure to take pictures on your field trips to use to enhance your presentation.

> **ROAD TRIP!**
>
> During the 2007–2008 challenge, our team brainstormed for weeks to find a building that was both unique and exciting. As a joke, one of the younger team members said, "Hey, we should go to the Mount Washington Observatory!" One of the older team members thought the idea was so cool that he called them and got permission. The whole team had a blast at Mount Washington, and the trip earned us top marks at our regional tournament.
>
> —Mindstorms Mayhem, New Hampshire

Discoveries and Inventions

Judges like seeing teams make new discoveries and inventions. Maybe you can think of a possible new application for a cutting-edge technology or a way to improve an existing application. Keep in mind that the Project includes finding a solution to a real-world problem related to your topic. This might consist of inventing a new application, such as a machine. But remember, you don't need to completely *finalize* an invention or idea; just having the idea is an advantage.

Don't rule out including in your presentation any prototype devices or possible software applications your team created during its research. Original research can be more difficult and time consuming, but judges won't overlook the hard work your team puts into doing its own experiments.

Present Your Research

Once you collect the *raw* information from the research, it's time to integrate it into a well-flowing, five-minute presentation. You want to clearly convey the information in an easy-to-understand way to the judges, and your team will need to share what it learned with the community to inspire others to become interested in science and technology.

Presentation Components

Presentations usually include at least three main components: information, discoveries, and examples of community outreach. Usually, you'll want to spend a good part of your presentation talking about the information you collected from research and your ideas like a solution to a problem. One good way to show examples of community outreach is to create a poster board or binder with pictures of your research, and show it to the judges when describing your work.

Presentation Style

FLL allows teams to give presentations in any style. Your style will affect your performance greatly, so choose wisely.

The following are some different ideas for presentation styles:

Skit   Don't forget your costumes, and practice your lines to perfection.

Song or poem   These are always entertaining, and it is impossible for a judge not to smile at them.

PowerPoint presentation   This is a standard presentation method and is still extremely useful, but avoid cramming too much info on each page.

Hands-On Activity   Involve the judges by having them participate in an experiment or act as assistants to the other team members.

"Panel of Experts"   Have one team member interview the other team members about the Project.

Storyboard   Instead of a video, create a storyboard, and have the team of "movie directors" explain how they will film the video.

Menu   Provide the judges with a restaurant-style menu, and have the team members (dressed as waiters) explain the various "dishes" that contain Project details.

When choosing a presentation style, pick one that involves the participation of as many team members as possible, such as a skit. A PowerPoint presentation can also involve a large number of members if they each give part of it.

NOTE   *If your team doesn't own a copy of PowerPoint, you can use the free Google Presentations tool, which has the same functionality. Find it by signing up at* http://docs.google.com/.

Also consider how well the team members present under pressure; some students get nervous when speaking to large groups or tournament judges. If some of your team members are nervous, they might want to do more "behind-the-scenes" support work, such as setting up laptops, poster boards, or other props used during the presentation. This allows them to participate in the presentation without actually presenting.

NOTE *FLL offers a great opportunity for nervous team members to learn to control nervousness and practice the valuable skill of speaking under pressure. One of the best ways to achieve this is through practice. Remember, the teams you will compete against are composed of other students. The judges will undoubtedly appreciate a valiant effort to do something that they can see is hard for team members.*

If you're considering a skit, opera, or any other presentation style that involves acting, consider how well your team members will perform. Some people can act out characters very well, which is a great advantage to this kind of style. However, other people find it hard to act naturally, resulting in a strained and unrealistic performance. If you have concerns about many of the members' acting skills, be careful.

Performance Tips

There are several good habits to observe during your presentation and the follow-up interview afterward. The following is a list of some:

- When talking to judges, make eye contact with them; don't look at the floor or at other team members. Remember to smile! Show the judges that you enjoy FLL and are enthusiastic about your Project.

- If your presentation involves interaction with the judges, consider involving them in some way. For example, if you're talking about an everyday product that could be made much less expensively when using a new technology, show the product to the judges, and ask how much they think it costs. Once they make some guesses, surprise them with the cheaper price.

- Project presentations are limited to just five minutes. Usually there is plenty to talk about, so make your points "short and sweet." Practice to refine your presentation.

- If you will use visuals in your presentation, be sure to use them at the right times and with appropriate references, rather than just flashing them up on the screen. Don't assume that your audience will follow along or understand what you're trying to communicate if you don't tell them.

- Avoid using presentation time to set up equipment. For example, if you will use a computer halfway through the presentation, set it up before you begin, or have one member set it up the while others give earlier parts of the presentation. Laptops are notorious for not working when you need them to. If you use a laptop, have it turned on and ready before you walk into the room of judges.

- If the members will talk, remember to talk at a normal speed and with your normal style, like you would in a regular conversation. Speaking too quickly or slowly can make the presentation sound unnatural or hard to understand.

- If your presentation is too long, you must make some hard decisions about what to cut. Keep in mind that you can bring up new information during the follow-up interview. For example, rather than discuss community outreach for 30 seconds during your presentation, you could talk about it during the follow-up interview instead.

Preparing

It's important to be well prepared for your presentation. When the members stumble over their parts, the presentation is hurt. The following are some ways to prepare for the presentation that will help to ensure that things run smoothly:

- Plan and outline your presentation. Rather than write down what you will say word for word, which could make the presentation sound unnatural, create a detailed outline and give your presentation using that. Have the members who will give the spoken parts of the presentation prepare their own outlines so that they can customize the pieces to fit their personalities.

- Once you outline the presentation, have the team members learn the pieces and practice. Each member should be able to deliver his or her piece smoothly, without needing to study his or her notes.

- Have coaches, parents, and mentors listen to the team practice and then comment on its performance in areas such as talking speed, posture, clarity, and so on.

- Have someone film the presentation, and then have the team review the performance.

Community Outreach

One portion of the Project involves sharing the team's research with the community. This can enable others to benefit from your work, inspire others to become interested in science and technology, and will be a great experience for the team. In addition, a Project's score is partly based on how much the team shares its research with others. This section discusses presenting your research to the community as well as taking other kinds of community action.

Unlike the presentation you give to judges, community presentations aren't limited by time or to a certain topic. Teams often expand community presentations to include information about the FLL competition itself and the robot. Some teams even bring the robot, field mat, and mission models to demonstrate what it can do. This can add a lot of excitement to the presentation, especially if the audience includes young children. Team videos and hands-on activities can also improve a presentation.

Be sure to take pictures of your community outreach, and find ways to include the pictures in your presentation to the judges. Perhaps include a Community Outreach *binder that documents the team's work in the community. Try to have a team member call local newspapers to inform them about upcoming community presentations; newspaper clippings that cover the event are a great way to share the team's community work with the judges.*

Many community organizations are happy to have FLL teams give presentations to members. For example, try one or more of the following:

- Boys & Girls Clubs
- Homeschool groups
- Museums
- Rotary clubs
- Schools
- Science organizations
- Senior living facilities
- Town hall meetings

You might even consider giving a presentation to the general public rather than a specific organization. For example, you could present in a building suitable for the occasion (such as a public library), and publicize the event with flyers and other advertisements. Figure 15-1 shows an actual flyer that a team used to advertise a presentation on its Project and robot.

Figure 15-1: A flyer used to advertise a team's presentation

Or talk to organizations about doing something based on the team's research. For example, a team that is researching the use of nanotechnology to create stain-resistant clothes might talk to a hospital about switching to nurses' uniforms made out of this material. Some teams go even further by working with politicians to pass legislation related to their research or even applying for patents for new inventions they create.

If you have an idea that is related to your research and can make a positive impact on your community, look for ways to make it happen! Though it might seem unlikely that organizations will listen to an FLL team, you may be surprised. Many organizations enjoy hearing what FLL teams have discovered and are open to suggestions for how they can use those ideas.

16

TOURNAMENTS AND BEYOND

Tournaments are one of the most exciting parts of
FLL. These fast-paced and fun-packed competitions
are the culmination of the teams' work on the Robot
Game and Project throughout the whole season.
Being prepared for a tournament can reduce the amount of stress involved,
especially if it's your first one. Although the specific program at each tournament varies, most have several basic events in common. This chapter gives
a rundown of these basic events and gives some tips to help prepare for them.
We also give ideas for activities you can do after the FLL season is over.

Pretournament Preparation

Before we get into discussing an actual tournament, you can do a few things
to help make sure you have everything you need.

Items to Bring

Although most of the items to bring to a tournament are obvious, you might not think of some other things that can come in handy, including the following:

- Tape measure
- Laptop (with robot programs) and USB download cable
- Field mat and mission models
- Extra pieces
- Extension cord(s)
- Power expansion strip

Tape Measure

A tape measure can help to demonstrate that your robot is under the height limit if referees ask. Keep a small one in your pocket or with other robot equipment just in case.

Laptop (with Robot Programs) and USB Download Cable

You will probably want to change some of the robot's programs at a tournament. For example, differences between the tournament's tables and environment and your practice table might cause the robot to perform differently, so be sure to come prepared to make tweaks to the robot's programs.

Field Mat and Mission Models

Bring your mat and models so you can practice with and test your robot without having to rely on the availability of the tournament's practice tables. You might also have a chance to show gracious professionalism to a team whose robot is in trouble by letting them use your equipment to test their robot.

Extra Pieces

Bring a generous assortment of extra pieces, including electrical parts such as sensors and motors, if possible. If you lose some of your robot's pieces, extra pieces can prove invaluable to your team or other teams if they lose pieces and didn't bring extras.

Extension Cord(s)

If you need electrical power at your pit area (discussed in "Set Up the Pit" on page 201) for things such as a laptop or battery charger, bring an extension cord, since close power outlets probably won't be available.

Power Expansion Strip

If you need multiple power outlets, be sure to bring a power strip. Again, having one of these might also enable you to help a neighboring team that needs power outlets.

Robot Packing and Storage

Due to the sensitivity of robots, make sure your robot is stored safely during the trip and at the tournament. Get a box to store the robot in whenever the team isn't using it. Just to be safe, you might want to put padding in the box to protect the robot during transportation. If you use an RIS robot, it's a good idea to cover the box with aluminum foil. This prevents nearby teams' programs from being inadvertently downloaded to your robot when it's on because the foil will deflect *IR beams* (beams of light that are used by the RIS to upload programs to the RCX). If you use attachments, store them in another designated box (or with the robot).

Use a Checklist

To make sure you take everything you need, you might use a checklist like the one that Figure 16-1 shows. A checklist can help you keep track of which items you still need to pack and make sure you don't forget anything. You may want to have the Equipment Manager (discussed in Chapter 6) take care of making and using a checklist.

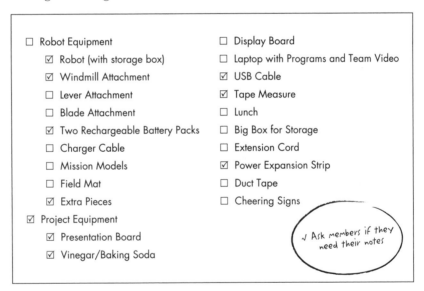

Figure 16-1: A sample checklist

Set Up the Pit

When you arrive at a tournament, you probably need to sign in first. You will usually receive a schedule of the tournament's events, including the times and places of your team's robot matches and judging interviews. You will also be assigned a table, called a *pit* area, where the members can store their things, display information about themselves, and so on. Judges sometimes interview teams at their pits as well.

Since your pit is the main way other teams learn about your team, consider decorating it with a poster showing information about your team or something similar. Prepare this poster before the tournament, including information such as your team name and number, pictures, sponsors, and so on. You might also want to create a display about your Project and/or set up a laptop to play a slideshow or video(s) of the team or robot.

If you keep valuables at the pit, such as a laptop or your robot, you may want to have at least one member (or coach, mentor, or parent) there most of the time to ensure that everything is safe. This also helps keep you from missing an impromptu interview with judges and enables you to be available if teams or tournament staff want to talk to you.

SPARE TIME

During the tournament, you will probably have a lot of "free" time when you don't need to attend an interview or compete in a robot match. Use this time for working on your robot or practicing your Project presentation (if you haven't done so yet). However, if you don't need to do anything, take some time to just have fun! Visit other teams or watch some of the robot matches. It can be a lot of fun and a learning experience to see how the other teams perform, how they solved some of the challenges, and what they did for their Project.

Judging Sessions

The Project, Robot Design, and Teamwork judging sessions usually occur early in the tournament but after teams have time to set up. When it's your turn, a tournament staff member may come to your pit to remind you about it and show you where to go. Each session usually takes about 15 minutes, and 2 to 4 judges score the teams.

General Performance Tips

It's important to make a good impression on the judges in each session. The best way to do this is probably to be friendly. Smile at the judges, make eye contact with them, and speak enthusiastically! After each session ends, be sure to thank the judges; they volunteered their time to make the tournament a reality and give you this great experience!

During the interview periods, try to have all the members answer questions to demonstrate how each member contributes to your team's efforts. If one member has already answered one or two questions, try to give other members a chance. You don't need an awkward silence to accomplish this; a member could simply ask another member if they'd like to answer a particular question. Younger members should especially try to answer any questions they can. Thinking of something important to add to someone's answer can be another way to contribute.

In all the sessions, coaches, mentors, or other adults should make sure not to participate in any way. Although adults are usually allowed to watch and film sessions, they aren't allowed to communicate with the team members or answer judges' questions. Any adult participation could result in a penalty to the team's score.

Be sure to check the Project, Robot Design, and Teamwork rubrics to get an idea of what the judges will look for in each session. You can find the rubrics at the back of the *FIRST LEGO League Coaches' Handbook*.

Project Session

You will probably have the Project judging session first. Your performance in this session, together with possible follow-up interviews at your pit, will determine your score for the Project component.

What Happens

When you first enter the room, the judges will ask for your team name, number, and perhaps some other team information. When the judges are ready, they start a timer and the team has a maximum of five minutes to give its presentation. Once you finish, the judges will ask questions about the Project. These questions might be about information in your presentation, how you researched certain things, or how the team worked together on various parts of the Project. During this interview section, you also have opportunities to tell the judges additional information, such as things you've done in the community.

Performance Tips

The *Coaches' Handbook* states that the five-minute time limit includes setup time. Once you enter the judging room, the judges can start the timer at any time, regardless of whether the team is ready or not. In reality, you'll probably have a reasonable amount of time in the room before you have to begin, but don't count on it. Ideally, you should be able to walk in and immediately start the presentation.

Try to find opportunities during the interview section to talk more about what the team has done with the Project. For example, you could talk more fully about a field trip you went on or a community presentation you gave. Demonstrate to the judges how much you did and learned!

Technical Interview

Your score for the Robot Design category is based mainly on your technical interview. Judges may also come by your pit area and ask you more questions if they think you're a candidate for a Robot Design award.

What Happens

In the technical interview, judges ask your team about your robot. This may include questions about how you designed and programmed it, how you worked together to solve challenges, and so on. There is usually a table with the mat and mission models set up so the team can demonstrate some of the robot's capabilities.

> ## A JUDGE'S PERSPECTIVE
>
> I've worked as an FLL technical judge many times. Typically when the judges first get together, we introduce ourselves and discuss our strategy for asking questions. Sometimes we agree to ask the same questions of each team, and other times we decide to let the team's presentation lead the questioning. What do I want to see and hear from a team? I like to see the excitement and sincere interest in your robot, its design, and its programs. I prefer to hear from every team member, no matter how small a contribution, and smiles and laughing are totally allowed. Don't be tense, and don't be nervous. The judges want to hear everything your team has to share and we really do want to help you relax and do your best. Make sure that every team member understands the robot's design and programs and can answer basic questions and you'll do fine.
>
> —JFK

Performance Tips

Don't be hesitant to talk about problems the team encountered with the robot. Judges love hearing how a team worked together to solve really tough problems. If you had a difficult situation but worked together to fixed it, be sure to bring it up during the interview.

You probably won't have enough time to demonstrate how the robot performs all the missions, and the judges aren't usually interested in seeing them all anyway. Instead, pick a few of the missions that demonstrate some of the robot's best features or that you solved in creative and unique ways, and show those missions to the judges.

Only the team members are allowed to do the work of building and programming the robot and solving missions, without help from adults. The judges will look to verify this and will want to see that the team understands how the robot's various mechanisms and programs work. Again, it is helpful if most or all of the members participate in describing the robot and its programs.

Teamwork Session

Similar to the Robot Design category, your score for the Teamwork category is determined mostly from the teamwork session. Judges may also watch your team during the rest of the tournament to see how the members interact with each other and with other teams.

What Happens

Teamwork sessions consist mostly of an interview in which the judges will ask about things such as team dynamics, work distribution, gracious professionalism, and the FLL experience. Some tournaments also include a small activity for each team to do during the session. As the members work on it, the judges observe their behavior toward one another and watch how they work together.

Performance Tips

Remember that the Teamwork session is about how well the team works together, not about how successful it was. Therefore, focus more on how to work through things and less on obtaining the best results (although it's certainly an advantage if you do well when working together!).

It's especially important to show how all the members contributed. Don't be hesitant to talk about problems the team worked together to solve. If your teamwork session includes an activity, go out of your way to include all the members.

Gracious professionalism is one of the core FLL values. It will be important to work together with others throughout your life, and the FLL competition is a good opportunity to begin learning that skill. Therefore, it's a good idea to be familiar with gracious professionalism and practice it throughout the season. You may even want to memorize the excerpt about gracious professionalism in the *FIRST LEGO League Coaches' Handbook* so you can demonstrate that you understand the idea well. Also mention any acts of gracious professionalism the team has done. For example, if you lend a LEGO piece to a team that loses one, mention it during the interview.

One possibly unexpected question we've noticed that judges frequently like to ask is how team members might react if they saw judges/referees arguing with a team or if they saw two teams arguing with each other. Think this through as a team so you know what you would do in one of those situations.

> ### COACHES' MEETING
>
> Before the Robot Game begins, many tournaments hold a *Coaches' Meeting* in which the Head Ref and other tournament staff review administrative items with the coaches from all teams. This may include last-minute rule changes, tournament guidelines, and so on, and it is also a good place to ask any questions about rules, procedures, or similar issues.

Robot Matches

The Robot Game usually begins after the judging sessions and opening ceremony. This is probably the most exciting and nerve-racking part of a tournament, when adrenaline is near its max.

How It Works

Tournaments choose the winners of the Robot Game in different ways. Each team has a minimum of three matches during the competition, with time in between each for recharging the robot and making any needed repairs and modifications. After these first matches, some tournaments may hold elimination rounds to determine the winners. Others determine the winners from the highest single scores in the first three matches. If the highest scores are tied, the second highest scores are used for a tie breaker, and so on.

The Match

Only two drivers are allowed at your table during a match. Depending on the tournament, you may be able to switch drivers during the match, but only two can be there at any given time. The rest of your team may be allowed to stand near the table to cheer and film, or they may not be allowed.

When you finish the match, a referee will review the missions with the team and determine your score. It's important to observe this process to make sure everything is scored correctly. Once the referee finishes and you check the score, you may need to sign the score sheet to signify your approval.

Between Matches

Depending on the size of the tournament, you can have as much as one to two hours between matches during the other teams' turns. Although this seems like a lot of time, you'll be surprised at how fast it goes by! This time is useful for recharging or replacing the robot's batteries and fixing any robot problems from the previous match. If the robot didn't perform correctly during a mission, hold a quick brainstorming session to determine the cause, and try to fix the problem before the next match.

If your robot uses a rechargeable battery pack or rechargeable batteries, bring backup batteries in case you don't have enough time to fully recharge the primary set before the next match.

Awards Ceremony

Once all the competition events are complete and the judges have had time to determine everyone's scores, the awards ceremony takes place. The teams are congratulated on their hard work, and the tournament gives out the awards. It's common for tournaments to recognize all teams during the awards ceremony and give them medallions for competing.

Teams may also receive notes from judges about their judging session performances. These notes usually list some of the strengths and weaknesses of the performance and may include comments from the judges. Keep these notes around; they can be useful for improving performance next year!

Figure 16-2: Awards from left to right: 1st Place Robot Performance Award (local Philadelphia tournament), 1st Place Champion's Award (local Philadelphia tournament), 1st Place Robot Performance Award (New Jersey State), 1st Place Champion's Award (New Jersey State), 1st Place Innovative Design (World Festival), 1st Place Robot Performance (World Festival)

Celebration

After a tournament ends, take some time to celebrate your accomplishments! This not only helps the team members feel good about the experience, but it's also a good opportunity for them to spend time with each other apart from simply working on the competition. To celebrate, some teams take a short trip, while others organize a get-together at a team member's house for a special dinner. The following is a list of some possible ways to celebrate:

- Amusement park
- Campout
- Pizza party
- Barbecue
- Customized cake (with congratulatory message and perhaps the team logo)

PAINTBALL CELEBRATION

After our state tournament in the Nano Quest season, my team held a celebration paintball game at our coach's house, topped off by a special custom cake brought by one of the team parents. All the members had a great time simply relaxing and having fun after a successful competition.

—JD

How to Handle Awards

Even though it's really exciting to win awards, you may have trouble deciding where to keep them, since all the members would probably like to have one.

Before choosing a permanent place for an award, pass it around between the members for a few months. One member (or the members of one family) could keep it for a couple weeks to show to friends and family and enjoy it, and they could pass it off to another member. Once an award circulates a few times, the members will probably be less anxious to keep it.

There are several ways to choose where to put awards, and the best way depends on each team's circumstances. If you win enough awards, you may be able to give one to each member or group of members that live close to one another. The members could pick numbers out of a hat to determine the order in which they pick their awards, for example.

If you only have one or two awards, it may be best to randomly pick the members that will get to keep them. Alternatively, the team might want to graciously give the awards to particular members, such as members who made exceptional contributions or members who will be too old to compete next season. A school team might keep its awards at the school instead of giving them to individual members.

Other Tournaments

If you don't make it past your qualifying or regional tournament, you might still be able to compete at official tournaments in other areas if you registered at the beginning of the season. Unofficial tournaments may not have the same restrictions. Use the tournament finder on the FLL website to find other nearby tournaments. You may also want to look at tournaments in nearby states or countries.

Before another tournament takes place, the team will probably have time to improve its performance. Here are some good ways to do this:

- Practice the Project presentation.
- Practice robot matches.
- Review performance(s) at previous tournament(s).
- Attempt new missions and/or refine current missions.
- Take more action in the community.

Practice the Project Presentation

Practicing is a great way to improve when the team doesn't have much time before the tournament. Have the members regularly run through the presentation to help stay "in shape" and perhaps get better. The team might also find ways to improve the presentation during practice.

Practice Robot Matches

Have the two robot drivers continue using the robot in practice matches. If the robot returns to Base several times for modifications, it's important for the drivers to handle the robot quickly and accurately, as this can increase the robot's consistency and decrease the time needed to attempt missions.

Review Performance(s) at Previous Tournament(s)

One of the best ways to prepare for another tournament is to look at your performance in the last tournament! Review the notes from your judging sessions (discussed in "Awards Ceremony" on page 206) to identify weaknesses, and then work on improving them. If a coach, mentor, or parent took videos of your judging sessions, carefully watch them to see ways you can improve things such as speaking style, posture, and so on.

If the robot encountered problems during the tournament, look for ways to keep those problems from happening in the next tournament. For example, the Robot Tables at the last tournament might have had bumpy field mats that gave your robot trouble. Try to find ways to make your robot less vulnerable to this problem in case it happens at the next tournament.

Attempt New Missions and/or Refine Current Missions

If you have enough time, you might want to design and program the robot to attempt more missions or modify it to perform current missions more consistently. You might think of completely new ways to attempt current missions that result in success more often or that reduce the time needed to attempt them.

If you make any modifications or additions to the robot's design or programs, be careful not to lose your previous work. It will be useful if you want to go back to it for some reason. Make sure the Program and Data Manager has documentation of the current robot and programs before beginning any modifications.

Take More Action in the Community

If you think it will help the Project, you might want to consider taking more community action. Although you might not have enough time or energy to take on any really ambitious projects, such as proposing legislation, you might do something simpler, such as giving another presentation to a group in your community. You could also try getting a newspaper or TV show to do a story on your team, especially if you did well at your previous tournament(s).

Robotics Workshops

Sometimes teams use the skills and experiences gained from an FLL season to start other activities. Some start robotics workshops to teach other students about LEGO robotics and FLL competition. Workshops are great ways to share knowledge with others and raise funds at the same time!

The following is a list of three common subjects for robotics workshops run by FLL teams. Workshops might incorporate any combination of them.

- LEGO Robot Basics
- Competing in FLL
- Team Recruitment

LEGO Robot Basics

Teaching the basics of building and programming LEGO robots is one of the most common subjects. Figure 16-3 shows an example of a syllabus for a week-long workshop that teaches the basics of LEGO robots.

```
LEGO MINDSTORMS Robot Basics

Day 1: You'll get an introduction to LEGO MINDSTORMS and its components. Then you'll
build and program a robot called TriBot from an instruction manual.

Day 2: You'll learn some basic building techniques. Then you'll build your very own robot,
using some of these techniques.

Day 3: You'll learn how to program basic commands with NXT-G. Then you'll program
your robot from Day 2 to perform various tasks.

Day 4: You'll learn about Touch and Sound Sensors and how to use them. Then you'll
add these sensors to your robot and program the robot to perform various tasks using them.

Day 5: You'll learn how to use Light and Ultrasonic Sensors, and will program your robot
to use them.
```

Figure 16-3: A sample syllabus for a workshop on LEGO robot basics

Competing in FLL

Since your team just went through the experience of competing in FLL, you might want to give a workshop to help others start their own teams. If new teams begin because of the workshop, you might offer further assistance by mentoring them or allowing them to watch how you run a couple of your meetings. Figure 16-4 shows an example syllabus for a four-day workshop on FLL.

```
Competing in FIRST LEGO League

Day 1: We'll give an introduction to FIRST LEGO League (FLL) and talk about its four
components and how they work. We'll also talk about starting and running a team.

Day 2: You'll learn how to compete in the Robot Game component of FLL. We'll give
techniques for building a successful robot and discuss strategies for attempting the missions.

Day 3: You'll learn about the Project component of FLL, including researching information,
creating a presentation, and reaching out to the community.

Day 4: We'll talk about the Robot Design and Teamwork components of FLL and give tips
on improving your performance in them.
```

Figure 16-4: A sample syllabus for a workshop about competing in FLL

Team Recruitment

Workshops can also be great ways to recruit new team members. Even in the first two kinds of workshops, you might meet other students who would be good additions to your team. Some teams give workshops specifically to find new members. In these workshops, you'd want to be able to see which participants would provide a useful contribution to your team. At the same time, you'd teach potential new members skills that would be helpful on your FLL team. Figure 16-5 shows an example syllabus for a two day-long workshop for team recruitment.

Team #735 "NXT Steppers" Recruitment

Day 1: We'll talk about our team, what we do, and why we're looking for new members. We'll also talk about what we're looking for in new members, such as skills in building, programming, research, public speaking, and teamwork. We'll also give some short lessons on building and programming LEGO MINDSTORMS robots.

Day 2: You'll participate in challenges that test your skills in the areas we discussed on Day 1. At the end of the day we'll invite participants who we feel would make good members to join the team. We'll also give tips to the other participants on forming their own teams.

Figure 16-5: A sample syllabus for a team recruitment workshop

Preparing for Next Season

While the team waits for next season, why not get a head start? Even though you won't know the missions and specific Project theme before the season starts, you can increase your general robot skills and become more knowledgeable in the area of the next challenge's theme. You may even be able to begin work on your robot!

Minicompetitions

Holding small-scale competitions is a fun way to increase members' skills in building and programming robot as well as research and presentation. Split the team into two or three groups, and give them a challenge with a deadline for completing it. The group that does the challenge best (in other words, the fastest, most efficiently, most creatively, etc.) is declared the winner. You could even give the winning members some small prize (such as candy).

The following is a sample robot challenge:

> Teams' robots start on opposite ends of a 6-by-10-foot table with 10 small, randomly placed "energy capsule" objects. When the timer is started, each robot has 1 minute to autonomously gather as many energy capsules as possible while staying on the table. The robots are allowed to interfere with each other's performance. At the end of 2 minutes, the robot that is touching the most capsules wins.

You can similarly organize Project competitions, except with a challenge related to research and/or presentation on the appropriate theme. Long before a season begins, its name is released along with a short description of the general theme. This information enables you to research the general area of the Project so the team will be more prepared when the details are released. The following is an example of a Project challenge that could have been held in preparation for the Nano Quest season:

> Each team must research nanotechnology and prepare a presentation on how it people use and develop it. Mr. Johnson and Mrs. Smith will watch each team's presentation and score them based on the amount of knowledge demonstrated, quality of research, and quality of presentation.

Building a Chassis and Bay

In Chapter 11, we talked about building a robot in three components: a chassis, bay, and attachments. Now we can see another advantage to using this method; the chassis and bay don't usually depend heavily on the specific details of each season's Robot Game. Therefore, you might be able to construct the entire chassis and bay before the season even begins! Of course, you might end up modifying both later, depending on the specifics of the challenge.

Some Final Thoughts

We hope this book has provided you with some useful information and advice on competing in FLL. Now it's up to you to do well and have fun in the competition. Put some good effort into your team, and enjoy yourself! And remember, FLL may just be the start of something much bigger for you. For example, you might find that you really enjoy designing robots or making discoveries in science and technology, and decide to pursue a career in those fields. The possibilities are endless!

RESOURCES

FLL teams can find a wealth of information on the Web related to the MINDSTORMS NXT robotics kit, as well as dozens of books on the subject. The following are some great resources to start with, but keep in mind that a good Google or Amazon.com search will turn up the most current resources.

Blogs

http://thenxtstep.com/ We're a little partial to this blog as we're both regular contributors. The blog provides news on all things related to MINDSTORMS NXT, including new products, building instructions, book reviews, and more.

http://www.nxtasy.org/ This is another useful blog that covers items related to MINDSTORMS NXT.

Forums

http://www.firstlegoleague.org/ FLL has an international forum for discussion about the competition. Look on its website to find the link. Note that you need to be a registered team to read and post in the forum.

http://messageboards.lego.com/   The LEGO Message Boards, a forum on the LEGO website, has a category for general discussion about MINDSTORMS.

http://thenxtstep.com/smf/   The NXT STEP blog also has a forum where visitors can post questions, upload pictures, and more. The blog's 30-plus contributors often respond to queries, and a dedicated forum section was created specifically for this book at *http://thenxtstep.com/smf/index.php?board=45.0/* (or just look in General Category ▶ Book Discussions).

http://www.nxtasy.org/   This blog's useful forum covers a wide range of technical topics for both beginner and expert.

Tutorials

http://www.ortop.org/NXT_Tutorial/   This site provides an excellent introduction to NXT-G programming. Once the basics have been mastered, advanced topics are also covered.

http://www.legoengineering.com/   This site has all sorts of resources and information related to NXT. It includes an online NXT Constructopedia that gives many useful tips on building with the NXT system, and it even has instructions for a couple of robots.

http://www.legoedwest.com/   LEGO Ed West has a collection of building instructions for both NXT and RCX robots, plus some articles on specific techniques such as using sensors.

Books

http://www.booksnbots.com/   At the Unofficial LEGO MINDSTORMS NXT Book Repository, author David J. Perdue (*The Unofficial LEGO MINDSTORMS NXT Inventor's Guide*, No Starch Press, 2007) keeps an up-to-date listing of books and workbooks related to MINDSTORMS NXT.

Online Stores

http://www.bricklink.com/ BrickLink is a huge online store, similar to eBay, except that only LEGO pieces and kits are sold. This site is a great place to find sets as well as single elements.

Random

http://mindstorms.lego.com/NXTLOG/ The MINDSTORMS' website has a feature called NXTLog, where people from all around the world post about their NXT projects. This site is a great place to find inspiration for robots or building and programming tips.

http://www.marsbasecommand.com/ Find fun activities (as well as good practice) for FLL teams to attempt before or after the season ends. The challenges mimic the FLL Robot Game by offering a playing field with mission models that a team's robot must attempt to manipulate.

http://www.nxtprograms.com/ This website includes an excellent collection of robot designs and programs for teams to try. Online building instructions are provided using clear photos of pieces. The NXT-G programs are available for download with comments to help decipher the function of the NXT-G blocks.

INDEX

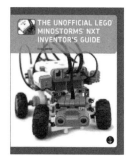

THE UNOFFICIAL LEGO® MINDSTORMS® NXT INVENTOR'S GUIDE

by DAVID J. PERDUE

The Unofficial LEGO MINDSTORMS NXT Inventor's Guide helps you to harness the capabilities of the NXT kit and effectively plan, build, and program your own NXT robots. After a brief introduction to the NXT, you'll explore the pieces in the kit and the roles they play in construction. Next, you'll learn practical building techniques, like how to build sturdy structures and work with gears. Then it's time to learn how to program with the official NXT-G programming language (and learn a bit about several unofficial programming languages, too). Finally, you'll follow step-by-step instructions for building, programming, and testing six robots (all of which can be built using only the parts found in one NXT kit).

OCTOBER 2007, 320 PP., $29.95
ISBN 978-1-59327-154-1

THE LEGO® MINDSTORMS® NXT ZOO!

An Unofficial, Kid-Friendly Guide to Building Animals with LEGO MINDSTORMS NXT

by FAY RHODES

Whether you're just beginning with your LEGO MINDSTORMS NXT set or are already an expert, you'll have hours of fun with these animal-like models that walk, crawl, hop, and roll! The first part of the book introduces you to the NXT kit and reviews the parts you'll need in order to begin building. Next, you'll learn how to program with the NXT-G programming language, including how to make miniprograms called My Blocks that you can use to build larger programs. Finally, you'll learn how to build each robot and program it to act like its real animal cousins. Learn to build and program models like Ribbit (a jumping frog), Bunny (a hopping rabbit), and Sandy (a walking camel).

FEBRUARY 2008, 336 PP., $24.95
ISBN 978-1-59327-170-1

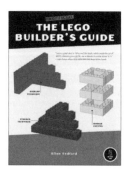

THE UNOFFICIAL LEGO® BUILDER'S GUIDE

by ALLAN BEDFORD

The Unofficial LEGO Builder's Guide combines techniques, principles, and reference information for building with LEGO bricks that go far beyond LEGO's official product instructions. Readers discover how to build everything from sturdy walls to a basic sphere, as well as projects including a mini space shuttle and a train station. The book also delves into advanced concepts such as scale and design. Includes essential terminology and the Brickopedia, a comprehensive guide to the different types of LEGO pieces.

SEPTEMBER 2005, 344 PP., $24.95
ISBN 978-1-59327-054-4

THE LEGO® MINDSTORMS® NXT IDEA BOOK

Design, Invent, and Build

by MARTIJN BOOGAARTS, JONATHAN A. DAUDELIN, BRIAN L. DAVIS,
JIM KELLY, DAVID LEVY, LOU MORRIS, FAY RHODES, RICK RHODES,
MATTHIAS PAUL SCHOLZ, CHRISTOPHER R. SMITH, *and* ROB TOROK;
foreword by CHRIS ANDERSON

The LEGO MINDSTORMS NXT Idea Book begins with an overview of the
NXT parts (beams, sensors, axles, gears, and so on) and explains how to
most effectively combine them to build and program working robots.
The book also delves into the complexities of the NXT programming
language (NXT-G) and offers tips for designing and programming robots,
using Bluetooth, creating an NXT remote control, troubleshooting, and
much more. Once you've mastered the basics, you'll get to show off your
skills building eight unique robots, including RaSPy (a robot that plays Rock,
Scissors, Paper), Slot Machine (complete with flashing lights and a lever),
and CraneBot (a crane-like grabbing robot).

SEPTEMBER 2007, 368 PP., $29.95
ISBN 978-1-59327-150-3

FORBIDDEN LEGO®

Build the Models Your Parents Warned You Against!

by ULRIK PILEGAARD *and* MIKE DOOLEY

Forbidden LEGO introduces you to the type of freestyle building that LEGO's
master builders do for fun in the back room. Using LEGO bricks in combination with common household materials (from rubber bands and glue to
plastic spoons and ping-pong balls) along with some very unorthodox building techniques, you'll learn to create working models that LEGO would
never endorse. Try your hand at a toy gun that shoots LEGO plates, a candy
catapult, a high-voltage LEGO vehicle, a continuous-fire ping-pong ball
launcher, and other incredibly fun inventions.

AUGUST 2007, 192 PP. *full color*, $24.95
ISBN 978-1-59327-137-4

PHONE:
800.420.7240 OR
415.863.9900
MONDAY THROUGH FRIDAY,
9 A.M. TO 5 P.M. (PST)

FAX:
415.863.9950
24 HOURS A DAY,
7 DAYS A WEEK

EMAIL:
SALES@NOSTARCH.COM

WEB:
WWW.NOSTARCH.COM

MAIL:
NO STARCH PRESS
555 DE HARO ST, SUITE 250
SAN FRANCISCO, CA 94107
USA

COLOPHON

The fonts used in *FIRST LEGO League* are Chevin, New Baskerville, Futura, and Dogma.

The book was printed and bound at Malloy Incorporated in Ann Arbor, Michigan. The paper is Glatfelter Spring Forge 60# Smooth, which is certified by the Sustainable Forestry Initiative (SFI). The book uses a RepKover binding, which allows it to lay flat when open.